Mental Math
for Pilots

Mental Math for Pilots

A Study Guide
by Ronald D. McElroy

Aviation Supplies & Academics, Inc.
Newcastle, Washington

Mental Math for Pilots
Second Edition
by Ronald D. McElroy
Cover Design and Layout by Laurel Johansen
Copyright 2004 by Ronald D. McElroy

Printed in the United States of America
Published 2004 by Aviation Supplies & Academics, Inc.
First Edition published 2000–2003 by Cage Consulting, Inc.

Aviation Supplies & Academics, Inc.
7005 132nd Place SE
Newcastle, WA 98059-3153
Email: asa@asa2fly.com
Internet: www.asa2fly.com

ASA-MATH-2
ISBN 1-56027-510-3

Library of Congress Catalog Card Number: 00-101041

Table of Contents

Introduction

There may be a myriad of reasons why you decided to take a look at this book—perhaps simply because of the title or cover. Regardless of your reason, I want this study guide to help any pilot sharpen his or her skills to better operate in the cockpit amid an ever-growing number of electronic gadgets designed to do our work for us. In short, because of the increased use of calculators and computers over the past few years, many of us either never learned, or have forgotten, the "tricks of the trade" that help us to work math problems. Whether it's simple addition or subtraction, multiplication or division, we have become increasingly reliant on electronics to enhance, supplement, or even replace some of our piloting skills. Forget about being able to do square roots or simple calculus in our heads!

So, what happens? We tend to get sloppy and over-reliant on the airplane "black boxes!" Many times we don't recognize errors quickly enough or even at all! The more advanced we become with our technology, the more mentally inefficient and lazy we become.

In this book we'll study the areas where pilots have traditionally needed to have sharp mental math skills. These include such subjects as fuel planning, temperature conversions, reciprocal headings, turn radius, crosswind components, time-speed-distance problems, calculating true airspeed, and the 60-to-1 rule, plus many others.

My goal in writing this study guide is to encourage and help you to be a more professional and precise pilot. As a result, you will be better armed to stay ahead of the flight by using the black boxes to assist you in planning the flight rather than being in the position of asking, "What's it doing now?" Or, for those of you without fancy computers to use inflight, this study guide will teach you many of the mental math tools and short cuts you will need to better fly and navigate. After all, the world of aviation is fast moving and multidimensional; we need all the help we can get just to fly from one airport to another.

Make a decision here and now to study and practice, practice, practice the mental math exercises discussed in this book. Once through—just scanning the exercises—won't do it for most people. Repetition is the key! Repetition is the key!

An additional benefit is in the area of career progression. Simply stated, this study guide may greatly improve your technical performance during each and every airline job interview you receive as you climb the ladder of an airline career. Don't underestimate the significance of this! With tens of thousands of qualified pilot applicants waiting for the chance, airlines can easily screen for the best of the best. So, include yourself in that category and be ready!

If airline interview preparation is your immediate goal, here are some suggestions:

1. Contact Cage Consulting, Inc. at 1-888-899-CAGE for professional airline interview preparation.

2. Build a personal study library to include: **AIM, FARs, ATP Test Prep, FEX Test Prep**, and the books *Checklist for Success: A Pilot's Guide to the Successful Airline Interview* and *Airline Pilot Technical Interviews*. Find everything you need to prepare at ASA's website, http://www.asa2fly.com.

3. Plan on 50 to 100 hours of study preparing for your interview. This may include technical study, a review of your own career, and administrative time preparing your application and reviewing your records. **Don't wait to prepare! Start now!**

I hope you enjoy my presentation of the material. But, as with *Airline Pilot Technical Interviews*, my success will be measured largely by the depth at which you are able to review and grasp the subjects discussed. I am sure you will learn something new that will help you fly the line a little better!

Ron McElroy

A Note From Cheryl Cage

In my dual roles as Pilot Career Consultant and author I am constantly on the lookout for book subjects that will help pilot applicants prepare for their interviews in the most efficient manner possible. I want my books to lower the stress level of each reader. I want to offer books that can be used for *every interview* a pilot attends on the road to reaching his or her career goals. I want the books of the Cage Consulting Series to be ragged-eared from use.

Little did I realize when Ron McElroy came to work as a Technical Consultant for Cage Consulting that I had also found the source of our third successful book *Airline Pilot Technical Interviews* (after *Checklist for Success*, the book and interactive CD).

Ron's aviation knowledge, his innate teaching abilities, coupled with a straightforward writing style made *Airline Pilot Technical Interviews* a best seller from day one. *Mental Math for Pilots* is another stellar performance from Ron McElroy. It fulfills all the requirements to be a Cage Consulting book. *Mental Math* will help raise your math skills thus lowering your stress level during an interview (or in the cockpit!). Whether you are preparing for your very first pilot job, have finally been invited to interview for your dream airline job, or simply are looking to enhance your math skills for everyday cockpit use, *Mental Math* **will** help you shine.

As always I love to hear from the readers of our books. You may email me directly at the address shown below.

Good luck!
Cheryl A. Cage
President
Cage Consulting, Inc.
cheryl@cageconsulting.com

Chapter 1

Taking the First Step

The root of mental math proficiency lies in the ability to grasp the basic concepts of addition, subtraction, multiplication, and division. The skill level you achieve is simply a reflection of how much work, or repetition, that you put into it. Starting now, in your everyday activities of paying for gas or groceries, giving an allowance to your kids, keeping track of sports statistics or scores, or calculating how much fuel to put in your aircraft, try to do all of them in your head. A good starting point is to write the numbers (or formulas) on a piece of paper, study how you solve the problem, and then push the paper aside and repeat the problem by visualizing what you have just completed. This takes a little extra time and discipline, but repetition and effort is as necessary here as it is with any other skill.

When you discover that you need to calculate a solution to a math problem, first define the problem; i.e., what is the answer you need? Second, look for the right formula to use. Most of the formulas you will ever need are right here in this book. Third, rearrange the formula to solve for the answer that you need. And, fourth, plug in the numbers and solve.

The same is true of the problems in this study guide. If you need to first complete the problems with pen and paper, do it! Once you've completed the problem, set the paper aside and repeat the problem in your head until you feel comfortable that you can repeat the solution in a timely manner without cheating.

Many of the subject areas in this book will have practice questions. The answers are in Chapter 7. In addition, there is a comprehensive test in Chapter 5 that will include different problems from all study areas. For all problems, try to be as accurate as possible. If you feel you need additional problems to solve, create some on your own. In fact, it will help increase your proficiency in solving problems to create your own problems.

I cannot overemphasize the importance and significance of having solid basic math skills. In most careers, having a slightly better-than-average skill will produce a noticeable increase in performance. That same philosophy is true in mental math skills for pilots. Therefore, Appendix A is available for extra study to review the basic concepts and techniques for solving simple and more complicated math problems we encounter while flying.

In Appendix A, the math skills that are reviewed include addition, subtraction, multiplication, division, squares, and square roots. In addition, there are problems to demonstrate simple and complex levels of proficiency, as well as practice problems for you to work.

The pilot population as a whole is no different in their math proficiency than in any other industry. Individual proficiency varies greatly. The remainder of this study guide relies on your ability to demonstrate basic math proficiency in order to understand and use the techniques and develop the skills necessary to increase your performance in the cockpit. Therefore, **I challenge you to review and assess your own math problem-solving skills, and make a commitment to study Appendix A.** If you are a new pilot your future employment may be at stake! If you are a seasoned veteran, your cockpit efficiency may improve significantly!

As a professional pilot you recognize the need to be at your top level of proficiency every time you fly. The preflight activities of flight planning, reviewing the weather, and checking the NOTAMs are a legal requirement of every flight. **However, the pilot skills you demonstrate are a reflection of the basic math skills and techniques you develop in a disciplined and focused strategy of study. This study guide is designed to help you be a more professional pilot!**

Study well. Good luck!

Chapter 2

Airborne Math Problems

Converting Hours and Minutes

I want to highlight the importance of the ability to convert hours and minutes as we normally read them from our watch into a more useful decimal value for use in math equations. Many pilots do not immediately recognize that one hour and fifteen minutes (1:15 or 1 + 15) **does not equal** 1.15 hours! Many have made the mistake of converting 1:15 to a decimal value of 1.15 (one point one five). The real answer is 1.25 hours!

So, here's the gouge. Since there are 60 minutes per hour, every 6 minutes is equal to one-tenth (0.1) hour. Thus, every multiple of 6 minutes is equal to the same multiple of tenths (0.1). And, if we desire greater accuracy, every 3 minutes is equal to one-twentieth (0.05) hour, or one-half the increment of 6 minutes.

You probably will not need more accuracy than a 3 minute interval in these conversions. Especially since it is the goal of this study guide to keep numbers and equations as simple and predictable as possible to allow reasonable mental math computations in the cockpit. However, just to be sure, I have constructed a short table of these intervals with their decimal equivalent for you to study and learn.

Minutes	Decimal Equivalent
3 minutes	0.05 hour
6 minutes	0.10 hour
9 minutes	0.15 hour
12 minutes	0.20 hour
15 minutes	0.25 hour
18 minutes	0.30 hour
21 minutes	0.35 hour
24 minutes	0.40 hour
27 minutes	0.45 hour
30 minutes	0.50 hour
33 minutes	0.55 hour
36 minutes	0.60 hour
39 minutes	0.65 hour
42 minutes	0.70 hour
45 minutes	0.75 hour
48 minutes	0.80 hour
51 minutes	0.85 hour
54 minutes	0.90 hour
57 minutes	0.95 hour
60 minutes	1.00 hour

Table 2-1

Reciprocal Headings

Seems a simple enough problem—yet, in the heat of battle you may freeze if you haven't practiced. Only two approaches are appropriate to get through this question: use a formula or visualize the headings on a compass rose.

Let's set up a practice table and work on using the formula.

Initial Heading	Reciprocal Heading
090°	270°
011°	191°
222°	042°
355°	175°
167°	347°
313°	133°

Table 2-2

Or, to look at it another way:

When INIT HDG <180°	INIT HDG + 200° – 20° = RECIP HDG°
When INIT HDG >180°	INIT HDG – 200° + 20° = RECIP HDG°

Table 2-3

Did you notice the change of the plus and minus signs between the formulas? We use two formulas because we will have initial headings either smaller than 180° or greater than 180° to begin the formula.

For example:

$$090° + 200° - 20° = 270°$$

Or

$$222° - 200° + 20° = 042°$$

Be cautious in using this formula for certain ranges of headings that will initially give you an answer that is either greater than 360° or less than zero in the first step of adding or subtracting 200. After completing the second step of adding or subtracting the 20, your answer will be corrected back into the appropriate range of 001° to 360°. Also, don't forget that the last digit always remains the same when computing the reciprocal.

For example:

$$167° + 200° - 20° = 347°$$

Or

$$191° - 200° + 20° = 011°$$

The second approach to figuring reciprocal headings—using the compass rose—comes simply with experience in flying on instruments. Study, visualize, and memorize the reciprocal cardinal compass headings. I recommend that you practice these reciprocals the next time you go fly. I believe you'll find it very productive and the more you work on it the easier it will become.

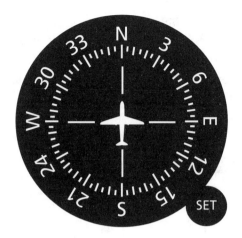

Illustration 1. Heading Indicator

Hydroplaning

It's always important to be aware of the potential for hydroplaning on wet runways. The popular convention for calculating hydroplaning speeds for your aircraft for either landing or takeoff is really quite easy. Aside from the hydroplaning formula, the only piece of information you need to have is your aircraft main tire pressure. The hydroplaning formula is simply calculated as:

$$V_{HP} = 9\sqrt{\text{Tire Pressure}}$$

As you see, knowing the square root of the tire pressure is important. Most high performance aircraft have a tire pressure in a range from, let's say, 80 psi to around 200 psi. That's quite a range, but if you approximate the values of the square roots at each end of the scale, you discover that the ranges are not really that wide. At 80 psi, the square root is about 9. At 200 psi, the square root is about 14. So, we now have a range of only 9 to 14. Next, in using the hydroplaning formula, you would multiply that value by a factor of 9. Thus, at the low end, 9 × 9 = 81 knots; and, at the high end, 9 × 14 = 126 knots.

You could, therefore, easily calculate your own specific aircraft hydroplaning speed that easily and quickly. Here are a few problems to test your skills. The answers are in Chapter 7.

Tire Pressure	V_{HP}
50 psi	?
120 psi	?
150 psi	?
230 psi	?

Table 2-4

Temperature Conversions

There are a few shortcuts to quickly convert Fahrenheit to Celsius and back again. It's important to memorize just a few important points, then use the tools to figure a rough estimate of the conversion. Here are the formulas, followed by a short table showing a couple of memorable temperature points:

$$F = (9/5 \times °C) + 32$$
$$C = (°F - 32) \times 5/9$$

Celsius	Fahrenheit
0°	32°
15°	59°
30°	86°
40°	104°

Table 2-5

Mental Math for Pilots

Some pilots are at ease using the conversion formulas. Perhaps they use them a lot; for me, I need a gimmick. Let me explain three other techniques.

Technique 1

First, if we note from the table above that the freezing temperature of water at 0°C equals 32°F, simply add or subtract 5°C for each 9°F or vice versa.

For example, let's figure what 30°C is in Fahrenheit. Remembering there are 5°C for each 9°F, 30°C is the same as (6 × 5°C) and 30°C is really 0° + 30°C. Now take the 6 (from 6 × 5°C) and multiply it by 9°F (6 × 9°F = 54) and add that result to 32°F. Remember 0°C = 32°F, getting 54°F + 32°F = 86°F.

To change from °F to °C, subtract 32 from °F and then do the multiplication. For example, 77°F − 32°F = 45°F, or 9 × 5. Multiplying 5 × 5 gives you 25°C.

It's relatively easy to use this 5 = 9 or 9 = 5 matching as long as you know just a few markers along the way. Try a couple of problems on your own, they're simple enough to catch on quickly.

Technique 2

The second technique to calculate °F is to double the °C, subtract 10%, and add 32. Or, to calculate °C, subtract 32 from the °F, add 10% and divide the result by 2. This is not very difficult and results in much more accuracy.

$$°F = ([°C \times 2] - 10\%) + 32$$
$$°C = ([°F - 32] + 10\%) \div 2$$

To use the previous example:

$$°F = 30°C \times 2 = 60°C$$
$$60°C - 6 = 54°C$$
$$54°C + 32 = 86°F$$

And

$$°C = 86°F - 32 = 54°F$$
$$54°F + 5 = 59°F$$
$$59°F \div 2 = 29.5°C \text{ (or pretty close to } 30°C)$$

Technique 3

The third way of estimating will get you in the ballpark for **lower** temperatures only. Either double the °C and add 30 to get °F, or subtract 30 from the °F and cut that in half to get °C.

$$°F = (2 \times °C) + 30$$
$$°C = (°F - 30) \div 2$$

An example:

$$2 \times 10°C = 20 + 30 = 50°F$$

And

$$50°F - 30 = 20 \div 2 = 10°C$$

(Note that if you use this method, for example, to convert 104°F to °C you will get 37°C, not the correct 40°C.)

Here are some practice problems for each of the techniques discussed (Table 2-6). When given a temperature in either °C or °F, convert to the other scale using each of the three techniques and compare the differences. A more detailed and complete temperature conversion chart is available for reference in Appendix B, Table 8.

°C	Technique # 1	Technique # 2	Technique # 3	°F
12°C	?	?	?	Calculate °F
25°C	?	?	?	Calculate °F
0°C	?	?	?	Calculate °F
Calculate °C	?	?	?	40°F
Calculate °C	?	?	?	81°F
Calculate °C	?	?	?	72°F

Table 2-6

Temperature Lapse Rate Deviations

The International Civil Aviation Organization (ICAO) has determined that the standard sea level temperature is 15°C, and that the standard temperature lapse rate is 2°C (or 3.5°F) per 1,000 feet change in altitude, up to 38,000 feet MSL. From this, you can determine deviations from the standard temperature for performance calculations during climb or cruise. The standard day temperature for each altitude is referred to with the term ISA (International Standard Atmosphere).

The moist adiabatic lapse rate is 2.5°C (4.5°F). Therefore, to estimate the possible cloud bases at an airport with a relatively close temperature/dewpoint spread, subtract the dewpoint from the actual temperature and then divide by the moist adiabatic lapse rate.

All of the practice problems use a straightforward method of solving for ISA temperature. Multiply the altitude, in thousands of feet MSL, by 2 (2°C temperature lapse rate); then, subtract from 15°C. For the first problem, 5 × 2 = 10; then 15°C minus 10°C equals 5°C, or the estimated standard temperature at 5,000 feet MSL. The rest of the problems are solved in the same manner. Compute the ISA temperature and deviation. To determine the standard deviation at those same altitudes, simply find the temperature difference between the actual and ISA temperatures.

Actual temp − ISA temp = temp deviation

Altitude	ISA Temp	Actual Temp	Temp Dev
5,000 MSL	?	20°C	?
8,000 MSL	?	15°C	?
FL 210	?	-10°C	?
FL 350	?	-60°C	?

Table 2-7

For reference, I have included a sampling of ISA temperatures versus altitude from sea level up through FL370 in Appendix B, Table 9.

What's the Pressure Altitude?

Sitting in the cockpit, if you set 29.92 in. Hg in your barometric altimeter, you would then be reading the standard day pressure altitude for your location. Simple enough, right?

However, a problem may occur when you must figure your pressure altitude based on a particular altimeter setting other than standard and using your local airport elevation.

This, again, is quite simple. For every .01 in. Hg altimeter setting, your pressure altitude reading changes 10 feet.

Q *The ATIS altimeter setting (QNH) is 29.79 in. Hg and the local airport elevation is 460' MSL. What is the pressure altitude?*

A *Pressure altitude equals 590'. The difference between 29.79 in. Hg and 29.92 in. Hg is .13 inches which converts to a difference 130' pressure altitude. Since we need to add the .13 in. Hg to 29.79 in. Hg to equal the standardized 29.92 in. Hg, we also add the 130' to the airport elevation of 460' to figure the pressure altitude.*

Q *As you are descending from FL350 for a landing, you forget to reset your altimeter to 30.57 in. Hg for the local airport QNH. What will your altimeter read after landing at the airport?*

A *The altimeter will read 650' low. The difference between QNE of 29.92 in. Hg and QNH of 30.57 in. Hg is .65 in. Hg which converts to a difference of 650' pressure altitude. After landing at the airport with a setting of 29.92 in. Hg, you would need to increase the barometric setting to 30.57 in. Hg to read the correct field elevation. Since the indicated altitude goes up as the barometric setting is increased, this means that at 29.92 in. Hg, the altimeter was reading 650' low!*

Crosswind Components

In position and holding for takeoff, or on short final for an approach, ATC gives you airport winds that seem a little strong and at a funny angle. How can you figure the crosswind component in a hurry to ensure you operate within the flight manual limitations?

Here's a quick technique I picked up that does an okay job of figuring a rough crosswind component. I'll demonstrate with the use of a small table followed by a detailed explanation. I have provided three ways of using the multiplier in the right column.

Please understand, these are rough estimates only and not necessarily mathematically exact. Close enough, though, for a quick estimate.

Wind Angle to Runway	Calculate Crosswind Component		
0 or 180	0.0	0%	None
030 or 150	0.5	50%	Half
045 or 135	0.7	70%	Two-thirds
060 or 120	0.9	90%	Almost all
090	1.0	100%	All

Table 2-8

To use this technique/table to approximate the crosswind component, first you have to determine the angular difference between the runway heading and the direction of the winds that ATC is providing you. Second, choose one of the closer "angles" from the table above. Third, multiply the total wind (including gusts) by the component from one of the right columns. Let's do a couple of quick problems. Just fill in the blanks.

Wind Angle to Runway	Total Wind Strength	Crosswind Component
030	20	?
050	20	?
070	18	?

Table 2-9A

As a quick check, using the table above, the answers I would have come up with are 10, 14, and 16 knots of crosswind component respectively. (Your answers could vary slightly if you rounded off differently than I did.) The key element here, for me, is to find a simple "crutch" that's easy to use in the cockpit when you are otherwise very busy.

Can you make the same crosswind calculation if the winds are between 090° and 180° offset from the runway? Sure. Using the same table, the crosswind components for 120°, 135°, and 150° offset match up with 060°, 045°, and 030°, respectively, on Table 2-8.

When you are determining the wind angle to the runway, compare reported magnetic winds to the actual magnetic heading of the runway. The **only** time you are ensured of getting magnetic winds is from the airfield tower controller and ATIS. All other reports or forecasts (METARs or TAFs) and PIREPs use true winds for reporting. At some airfields, the magnetic variation can be large enough to make a significant difference in computing wind components.

Another great technique for computing crosswind components is to first determine the wind angle to the runway. Second, add a value of twenty (20) to the wind angle, the total of which is now to be used as a percentage for the next step. Third, multiply the total wind value by the percentage from Step 2. The result is a value for the crosswind component.

For example, if there is a wind on runway 25 reported as 280/18, what is the crosswind component?

1. The wind angle to the runway is 30 degrees.

2. Add the number 20 to 30; e.g., 20 + 30 = 50%

3. Multiply the wind value by the percentage; e.g., 18 × 50% = 9 knots crosswind.

Try this technique on the problems in Table 2-9A to determine if this technique compares closely to the other techniques described.

Headwind and Tailwind Components

The discussion and techniques used for computing crosswind components can be modified and used in a similar manner for computing headwind or tailwind components. The simplified math tools to use in multiplying the reported wind velocity are presented in Table 2-9B. Please understand, these are rough estimates only and not necessarily mathematically exact. Close enough, though, for a quick estimate and ease of mental math calculations.

Wind Angle to Runway	Calculated Headwind or Tailwind Component		
0 or 180	1.0	100%	All
030 or 150	0.9	90%	Almost all
045 or 135	0.7	70%	Two-thirds
060 or 120	0.5	50%	Half
090	0.0	0%	None

Table 2-9B

Notice that the table is "inverted" from the cross-wind component table illustrated in Table 2-8. Mathematically, this is due to the relationship of angles and the properties of the sine and cosine of right-angle triangles.

Remember the **Pythagorean Theorem** from your academic days? This stated that in a right triangle (with a 90-degree angle in one corner) when you square the length of each adjacent side to the 90-degree angle (side A and side B), the sum of those two sides when squared is equal to the length of the third side of the triangle (also called the hypotenuse, side C) when squared.

$$A^2 + B^2 = C^2$$

The reason for remembering this relationship is to know that when you add each crosswind and headwind component, the sum of each component will be greater than the value for the total wind component. That is the basis for the Pythagorean Theorem. In other words, a 6-knot crosswind plus an 8-knot headwind adds up to a larger value (14) than the actual 10-knot total wind. However, $6^2 + 8^2 = 10^2$, or $36 + 64 = 100$ is correct, which supports the Pythagorean Theorem for right triangles.

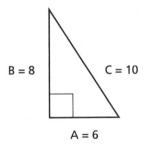

B = 8   C = 10

A = 6

In review, how would you determine that you have exceeded your 10-knot tailwind limitation for takeoff or landing? Using only the tailwind multipliers from Table 2-9B, I've illustrated total wind values from different selected wind angles that will result in an approximate 10-knot tailwind component. Don't forget to add the gust value to the total wind value to determine the actual tailwind component (which also applies to calculating the headwind or crosswind components).

Wind Angle to Runway	Total Wind Value	Component Multiplier	Resultant Tailwind Value
180	10 knots	100%	10 knots
150	11 knots	90%	10 knots
135	14 knots	70%	10 knots
120	20 knots	50%	10 knots
090	Any speed	0%	0 knots

Table 2-9C

Crosswinds versus Drift Angle

Have you ever tried to estimate what your drift angle would be when at cruise or on the approach? Here's a formula to make a quick estimate. Note that if the crosswinds become exceedingly large, this formula will not be as accurate for drift angles greater than 15 degrees.

Drift Angle =
    [(Crosswind Component) × 60] ÷ TAS

For example, on approach with a 120 KTAS, you have a 20-knot crosswind component during the approach and landing. What is your expected drift angle?

Drift Angle = (20 × 60) ÷ 120 =
  (1200) ÷ 120 = 10°

For another example, at cruise at 420 KTAS, you have an 84-knot crosswind component. What is your expected drift angle?

Drift Angle = (84 × 60) ÷ 420 =
  5040 ÷ 420 = 12°

This formula can produce very large numbers during the multiplication. So, now that you have a better understanding of the formula, consider the next technique as a much easier mental math technique for computing crosswinds versus drift angle.

A mental math crutch for computing drift angle is to remember that for a crosswind component equal to your speed expressed in miles per minute (60 KTAS equals 1 NM per minute), you will have one degree of drift angle. In addition, if your speed is expressed in Mach number (which is a percentage of the speed of sound), for a crosswind component equal to your Mach number × 10, you will have one degree of drift angle. As an example, if you are cruising at .80 Mach (approximately 480 KTAS or 8 NM per minute), you will have one degree of crab for each 8 knots of crosswind component.

Here are a few more practice problems.

TAS	Crosswind Component	Drift Angle
150 knots	12 knots	?
360 knots	48 knots	?
90 knots	30 knots	?
0.78 Mach	50 knots	?

Table 2-10

Conversion Factors

We often have a need to convert miles per hour to knots, statute miles to nautical miles to kilometers, etc. Here is a short conversion table to remind you of those conversion factors to make the transition between the varying units of measure. Next is a table defining the units of measurement.

1 Statute Mile = 0.87 Nautical Mile (NM)	1 Nautical Mile = 1.15 Statute Mile (SM)
1 MPH = 0.87 Knot	1 Knot = 1.15 MPH
1 Statute Mile = 1.61 Kilometers (KM)	1 KM = 0.62 Statute Miles
1 MPH = 1.61 Kilometer Per Hour (KPH)	1 KPH = 0.62 Miles Per Hour (MPH)
1 NM = 1.85 KM	1 KM = 0.54 NM
1 Knot = 0.51 Meters Per Sec (MPS)	1 MPS = 2 Knots
1 Knot = 1.85 KPH = 0.51 MPS	1 KPH = 0.54 Knots

Table 2-11

Meters	Units of Measurement	Feet
1,609	Statute Mile (SM)	5,280
1,852	Nautical Mile (NM)	6,076
1,000	Kilometer (KM)	3,208

Table 2-12

I hope that you also want to keep these conversions as simple and easy to remember as I do. Therefore, just approximate! That's right, simply remember a number that is close; e.g., 1 knot is 15 percent more than 1 MPH, 1 MPS is about 2 knots, 1 SM is a little more than 1½ KM, etc. Now, here's a couple of practice problems. And, remember, keep it simple!

Given:	Find:
200 Knots	? MPH
180 MPH	? Knots
8 MPS	? Knots
9 KM	? SM

Table 2-13

Visibility to RVR Conversions

Remember that visibility is given in statute miles and runway visual range is given in feet. In addition, visibilities-to-RVR conversions do not have a linear relationship. My suggestion is that you memorize this short conversion table.

Visibility	RVR
¼ Statute Mile	1,600 feet
½ Statute Mile	2,400 feet
¾ Statute Mile	4,000 feet
1 Statute Mile	5,000 feet
1¼ Statute Mile	6,000 feet

Table 2-14

Fuel Planning

Fuel planning can be important for three reasons:
• Do I have enough fuel to be legal for my flight?
• How much fuel do I need to upload for my flight?
• Is my fuel burn in flight consistent with my flight planning?

Let's look at the requirements for fuel planning based on FAR §§91.151 (VFR) and 91.167 (IFR). Assume the VFR flight is planned for 2 hours and 20 minutes. Assume the IFR flight is planned for 3 hours and 15 minutes plus 40 minutes to the alternate.

VFR Rules		IFR Rules
2 + 20	ETE	3 + 15
Not Required	Alternate	0 + 40
0 + 30 or 0 + 45	Reserve – Day or Reserve – Night	0 + 45 or 0 + 45
2 + 50 (Day) 3 + 05 (Night)	Total Fuel Required	4 + 40

Table 2-15

Once you determine the amount of hours of fuel you need to start the flight, convert the hours to either gallons or pounds of fuel, depending on the measurement needed for your flight operations. What's the conversion formula?

Avgas:

Total pounds Avgas =
(# Gallons) × (6.0 lbs per gallon)

Total gallons Avgas =
(Pounds Avgas) ÷ (6.0 lbs per gallon),

or a mental math shortcut:

Total gallons Avgas =
[(Pounds Avgas) × ($1\frac{2}{3}$)] ÷ 10

Jet A:

Total pounds Jet A =
    (# Gallons) × (6.7 lbs per gallon)

Total gallons Jet A =
    (Pounds Jet A) ÷ (6.7 lbs per gallon),

or a mental math shortcut:

Total gallons Jet A =
    [(Pounds Jet A) × ($1\frac{1}{2}$)] ÷ 10

Using these conversion formulas, fill in the blanks of the following practice problems.

Gallons	Pounds
55 gal. Avgas	?
?	480 lbs. Avgas
?	1,000 lbs. Avgas
500 gal. Jet A	?
?	5,000 lbs. Jet A
?	8,500 lbs. Jet A

Table 2-16

Fuel Dumping

These tend to be easier problems to work in your head because there are usually only three variables (dump rate, time, fuel dumped) to work within the formula. Two of the three will be given to you, leaving you to determine the third variable.

The greater challenge in this problem is working with large numbers, probably in the thousands of pounds. But don't get too anxious, the problems are normally designed so that the answers work out in round numbers. And, better yet, many of the questions I've heard of asked during job interviews have you solve only for the time variable. Therefore, your practice of these problems can be more methodical and consistent.

(Fuel Dumped) ÷ (Dump Rate) = (Time)

or

(Dump Rate) × (Time) = (Fuel Dumped)

Here's a table of problems for you to practice with. The answers can be found in Chapter 7.

Dump Rate	Time	Fuel Dumped
1,300 PPM	?	6,500 lbs
2,500 PPM	?	45,000 lbs
3,000 PPM	?	19,000 lbs
2,500 PPM	?	30,000 lbs
2,200 PPM	?	11,000 lbs
1,500 PPM	7 min.	?
1,200 PPM	11 min.	?
?	5 min.	12,500 lbs
?	16 min.	48,000 lbs
2,000 PPM	?	20,000 lbs

Table 2-17

It will be much simpler if you remember to drop two zeroes from the end of each number just to keep the numbers more manageable in size rather than outrageously large.

Next, use one of two approaches to divide the Fuel Dumped by the Dump Rate. Using the first technique on the first problem, you would divide 65 by 13 (after dropping the last two zeroes) which equals 5 minutes. This method is strictly a mathematical approach that some can readily calculate in their head.

The other technique is to use a method of proportions to arrive at the proper solution. In the second problem, using the numbers 25 and 450, I would first double the dump rate to 50 so that I could more easily recognize that $450 \div 50 = 9$; therefore, $450 \div 25$ (the same as $50 \div 2$) = 18 (or 9×2) minutes.

Let's look again at the first problem using a variation on this technique. Using the numbers 13 and 65, I first double 13 to get 26 (a multiplier of 2). I then double 26 to get 52 (now a multiplier of 4). Then I recognize that I have a remainder of 13 (a multiplier of 1). Thus, I now have multipliers of 4 plus 1, which equals 5. This is the correct answer.

Let's do this again using the third problem, using the numbers 3 and 19 (I can drop three zeroes in each number). I know that if I multiply 3 by 6 the result is 18 with a remainder of 1. What do I do with this? So far we know the answer is 6 minutes plus something. But, to be exact, we can see that the remainder of 1 divided by the dump rate of 3 equals one-third of a minute. Therefore, we now have an exact answer of 6 minutes and 20 seconds.

Magnetic Compass Turns

Using the magnetic compass in the cockpit as the sole reference for turns requires a special awareness of the following magnetic compass characteristics which are caused by magnetic dip.

Note: These characteristics are only applicable in the Northern Hemisphere. In the Southern Hemisphere the characteristics are observed in the opposite direction.

If a turn is made to a northerly heading from any direction, the compass indication when approaching north lags behind the turn. Therefore, the rollout of the turn is made before the desired heading is reached. If a turn is made to a southerly heading from any direction, the compass indication when approaching southerly headings leads ahead of the turn. Therefore, the rollout is made after the desired heading is passed. The amount of lead or lag is at a maximum on the north–south headings and depends upon the angle of bank used and the latitude of the airplane. The following acronym may help remember these characteristics.

UNOS — Undershoot North, Overshoot South

In addition, when on an east or west heading, an increase in airspeed or acceleration will cause the compass to indicate a turn toward north. A decrease in airspeed or deceleration will cause the compass to indicate a turn toward south. Use this acronym to remember these characteristics.

ANDS — Accelerate North, Decelerate South

If we assume that the lead point for rolling out of a turn is normally $\frac{1}{3}$ of the bank angle, let's calculate the revised lead point using the magnetic compass only for the turn and rollout.

Bank Angle (Left/Right)	Start Heading	Desired Heading	Latitude	Lead Point
15° R	270°	360°	30° North	325°
15° L	270°	180°	30° North	155°
25° L	090°	010°	40° North	058°
25° R	090°	190°	40° North	222°

Table 2-18

Mental Math for Pilots

Using the UNOS acronym, let's look at the solution for these problems. The first problem normally uses a lead point of 5° (15 ÷ 3) plus the latitude of 30 degrees north, for a new revised lead point of 5 plus 30, or 35°. With the desired heading of 360°, a 35° lead point results in rolling out of the turn when the magnetic compass reads 325°. The rest of the examples are solved the exact same way. Here are a few practice problems; but, there is one problem that may trick you. Just stick with the characteristics as explained above.

Bank Angle (Left/ Right)	Start Heading	Desired Heading	Latitude	Lead Point
15° R	270°	360°	45° North	?
15° L	270°	180°	34° North	?
25° R	360°	090°	40° North	?
20° R	090°	190°	40° North	?

Table 2-19

60-to-1 Rule

Remember this from your basic instrument course way back when? Well, I don't know specifically that the majority of pilots ever use this rule, but, occasionally an understanding of this concept will help you solve problems related to course or DME intercepts during departure, arrival, or approach procedures. Here we go.

The 60-to-1 rule means that at 60 DME from a VOR every 1° of course deviation equals 1 NM (approximately 6,000'). Let me illustrate this in three formats: a table, a formula, and an illustration.

DME from VOR	1 Degree = ? NM
60	1
30	½ or .50
20	⅓ or .33
15	¼ or .25
12	⅕ or .20
10	⅙ or .16

Table 2-20

# of Radials per mile = 60 ÷ DME

or

Width of 1° (NM) = DME ÷ 60

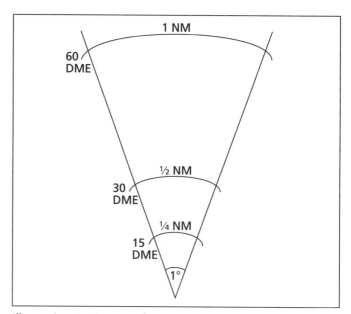

Illustration 2. 60-to-1 Rule

Pictorially, this 1 degree slice of pie in the sky looks like the previous diagram. Perhaps it will better illustrate the relationships of the DME to the distance per degree.

Q *On the ILS 19R approach plate for XYZ airport, there is a 15 DME arc transition from the XYZ 047/15 (Sparky) to the XYZ 011/15 (Falfa). What is the distance along the arc from Sparky to Falfa?*

A 9 NM.

Step one: Calculate how many radials are crossed.

047° minus 011° = 36 radials

Step two: Calculate the number of radials per nautical mile.

60 ÷ 15 DME = 4 radials per NM

Step three: Combine the answers from above.

36 radials ÷ 4 radials per NM = 9 NM

One suggestion I have for mental math simplification is to try to use DMEs that make the formulas work out easily; i.e., if you are given a 16 DME arc, use 15 DME instead. (**Note:** But don't fly it! Fly the 16 mile arc!) Here's another problem to solve.

Q *When flying the same ILS 19R transition using the 15 NM arc, how many degrees "lead" would you need to start the turn from the arc to the localizer, using a standard rate turn (SRT) at 200 knots? (Hint: SRT at 200 knots equals 30 degrees bank and 1 NM turn radius.) In addition, work two additional problems using the 20 NM and 12 NM arc transition to the localizer. (The answers are in Chapter 7.)*

DME Arc	Lead Radials
15 NM	? degrees
20 NM	? degrees
12 NM	? degrees

Table 2-21

Bank Angle for Standard Rate Turn

The bank angle for a standard rate turn can be estimated by a simple formula or looking at the Turning Performance Chart in Appendix B, Table 7. SRT is defined as turning at the rate of 3 degrees per second. The formula for a standard rate turn bank angle is:

Bank Angle [SRT] = (TAS ÷ 10) × 1.5

However, for IFR operations, normally a maximum of 30 degrees of bank is used even if the standard turn rate of 3 degrees per second is not achieved. Therefore, for any true airspeed greater than 200 knots, the maximum bank angle used would be 30 degrees. This limitation is used for computing turn radii in the practice problems.

Turn Radius

Turn radius calculations are very simple using either of two techniques. Both techniques assume that the required turn radius equals the turn lead point distance when the number of degrees to be turned is 90 degrees. The turn lead point distance will be progressively less when the turn angle is less than 90 degrees. Both techniques also assume that the bank angle for a SRT is used.

The first technique seems to be more accurate at lower speed; i.e., in the landing pattern at 200 KIAS and below, using a bank angle for a standard rate turn. Here's the formula:

Turn Radius [NM] = [TAS] ÷ 200

Or

$$\text{Turn Radius } [(\text{NM}] = \frac{\text{TAS} \times 1\%}{2}$$

The next technique seems to be more accurate at higher speed; i.e., at cruise flight levels using indicated Mach numbers approximately 0.40M and greater, or at 200 KIAS and greater, and using a bank angle of 30 degrees. Here's the high-speed formula:

Turn Radius [NM] = (Mach Number × 10) – 2

KTAS	Standard Rate Turn Bank Angle	Turn Radius (SRT/30° Bank max)
90 KTAS	? degrees	? NM
120 KTAS	? degrees	? NM
200 KTAS	? degrees	? NM
0.80 Mach	30 degrees (max IFR)	? NM

Table 2-22

The first practice problem is solved as follows; the remainder of the answers are in Chapter 7.

Step one: The SRT bank angle is calculated to be 90 ÷ 10 = 9; then 1.5 × 9 ≈ 14° degrees bank at SRT.

Step two: The turn radius is calculated by taking one percent of 90, which equals 0.9; then dividing by 2. The final answer is 0.45 or approximately 0.5 NM turn radius.

Calculating True Airspeed

To calculate your true airspeed mentally, simply add 2% per thousand feet above mean sea level to your indicated airspeed. There will be some variation at higher altitudes and with temperature deviations, but this formula is normally adequate for a quick check. Here is the formula and a few practice problems.

$$KTAS = KIAS + (KIAS \times [\text{Altitude \{in 1000s\}} \times 2\%])$$

KIAS	Altitude	KTAS
100 KIAS	10,000 ft MSL	?
140 KIAS	5,000 ft MSL	?
200 KIAS	13,000 ft MSL	?
280 KIAS	FL350	?

Table 2-23

Time-Speed-Distance Problems

This is one subject I've noticed that seems to challenge pilots the most. I believe that the reason is simply neglect rather than ignorance. Perhaps I can knock the dust out of the inner chambers with a few new mental math techniques.

First, remember that there will be only four variables in the formula, and three of them will always be known. The four variables are **wind**, **distance**, **time**, and **true airspeed**.

Second, I've found it helpful, after hearing the three numbers, to immediately convert the speed (if given) into miles per minute. Say again? That's right, it seems to be easier to solve most problems when you know how many miles per minute you are traveling.

If you don't have a good grasp of converting knots to miles per minute, study this chart and memorize the relationships, even the half-mile per minute numbers. As an additional crutch, remember every 30 knots equals $\frac{1}{2}$ mile per minute, or 60 knots equals 1 mile per minute. You can then quickly figure that, for example, 480 knots equals 8 miles per minute.

Ground Speed	Miles per Min
90	1.5
120	2
150	2.5
180	3
210	3.5
240	4
270	4.5
300	5
330	5.5
360	6
390	6.5
420	7
450	7.5
480	8
510	8.5

Table 2-24

The basic time-speed-distance formula is this:

(GS) × (Time) = (Distance)

The speed variable in the formula refers to ground speed. Therefore, **you must add or subtract the headwind or tailwind component to the true** airspeed to get the ground speed in knots; i.e., TAS ± Wind = GS.

Here's a table of problems containing four variables. Three are given to you, leaving the fourth for you to solve. Answers will be shown in Chapter 7.

KTAS	Wind	Time	Distance
240	60 TW	?	200 NM
280	70 HW	10 min.	?
150	0	?	5 NM
?	0	4 min.	20 NM
420	60 TW	?	400 NM
?	0	2 min.	14 NM
?	0	1.5 hr.	600 NM
500	0	45 min.	?
?	0	40 min.	340 NM

Table 2-25

Here's how to use my technique of converting ground speed to miles per minute to make the problem solving easier without use of a calculator or pad and pencil to work the formula.

As soon as you are given the ground speed, or can figure it out from the true airspeed and winds, convert that to miles per minute. Example: 350 knots GS equals 360 ÷ 60 = 6 MPM (notice that 350 is close enough to 360 to use since it is an exact multiple of 60).

You can then more easily multiply this miles per minute figure by the number of minutes to get the distance traveled. Or, you can divide the distance by the miles per minute to figure the number of minutes. For example: 10 minutes @ 6 MPM = 60 miles. Or, 90 miles @ 6 MPM = 15 minutes.

Notice the last three problems in Table 2-25 can be more easily solved by using an approach of proportions. That is, if you realize that 1.5 hours is three segments of 0.5 hours, and that 600 NM is three segments of 200 NM, then you realize that you travel 200 NM per half-hour or 400 NM per hour. Thus, 400 knots ground speed.

Next, realizing 45 minutes is three segments of 15 minutes, and that 500 knots is a distance of 250 NM each 30 minutes or 125 NM each 15 minutes, then, three segments of the 125 NM per 15 minutes equals 375 NM in 45 minutes.

Finally, in the last problem, one easy approach to figuring the GS is to recognize that since you've traveled 340 NM in 40 minutes you would also travel 170 NM in 20 (divide both variables by 2). Then, multiply both of these by a factor of 3 which results in traveling 510 NM (170 × 3) in 60 minutes (20 × 3) or one hour.

Some people find it helpful to verbalize their problem solving out loud. I agree with this technique, **if you believe this will help you to be more methodical**. Be careful, however, since any blundering of the numbers out loud could be quite embarrassing! Having said this, could you repeat solving the problems in the table with a little more ease and confidence? If not, keep trying until you can. It's worth the effort! Here's another word problem to test your skill.

Q *You are cruising at FL310 at 480 knots ground speed. You have been cleared to descend to 13,000' MSL, and you know your aircraft will average 3,000 fpm during the descent. How far will you travel during the descent?*

A *48 NM.*

Here's how I would set it up to solve the problem in my head.

Step one: My speed at 480 knots is 8 nautical miles per minute.

Step two: I need to descend 18,000 feet at an average rate of 3,000 fpm which gives me 6 minutes to descend.

Step three: To find the distance traveled, I multiply (8 NM/min.) × (6 min.) which equals 48 NM.

Chapter 3

Calculating Enroute Descents

There are several ways to calculate enroute descents. Often, however, pilots become confused because of the many techniques available. Once you are comfortable with one method for these enroute descent calculations in your current aircraft, you will be able to perform the calculations quickly and accurately. However, for now I will assume that you have not yet settled on a particular method.

There are three basic methods to calculate enroute descents: the 3-to-1 rule, the constant descent rate, and the pitch attitude solution. Each method has a unique advantage for specific types of operations. As we discuss them, I'll make a recommendation for pairing a method with a type of operation or type of aircraft.

(MOD.MOD2) 95201
MODESTO TWO ARRIVAL   ST-375 (FAA)

SAN FRANCISCO INTL
SAN FRANCISCO, CALIFORNIA

OAKLAND CENTER
134.37 281.5
BAY APP CON
134.5 336.2
ATIS
113.7 118.85

MUSTANG
117.9 FMG
Chan 126
N39°31.88'-W119°39.36'
L-5-7, H-2

MANTECA
116.0 ECA
Chan 107

MINA
115.1 MVA
Chan 98
N38°33.92'-W118°01.97'
L-3, H-2

SCAGGS ISLAND
112.1 SGD
Chan 58

FL 220
182°
(115)

INYOE
N37°53.74'
W118°45.90'

MODESTO
114.6 MOD
Chan 93
N37°37.64'-W120°57.47'

N37°49.56'
W119°22.09'

FL 190
204°
(53)

OAKLAND
116.8 OAK
Chan 115

FL 180
244°
(50)

FL 190
246°
(48)

R-215

R-183

R-107

R-093

R-064

8000
245°
(16)

(16)

FL 180
244°
(27)

COALDALE
117.7 OAL
Chan 124
N38°00.20'-W117°46.23'
L-3, H-2

TROSE
N37°41.95'
W120°24.25'

GROAN
N37°35.42'
W121°17.11'

273°
(26)

CEDES
N37°33.05'
W121°37.48'

N37°28.56'
W120°26.39'

FAITH
N37°24.07'
W121°51.71'

Expect clearance to cross at
11,000 and 250K IAS.

SAN FRANCISCO
115.8 SFO
Chan 105

8000
305°
(47)

CLOVIS
112.9 CZQ
Chan 76
N36°53.06'-W119°48.91'
L-2, H-2

NOTE: Chart not to scale.

CLOVIS TRANSITION (CZQ.MOD2): From over CZQ VORTAC via CZQ R-305 and
MOD R-093 to MOD VOR/DME. Thence . . . .
COALDALE TRANSITION (OAL.MOD2): From over OAL VORTAC via OAL R-246
and MOD R-064 to MOD VOR/DME. Thence . . . .
MINA TRANSITION (MVA.MOD2): From over MVA VORTAC via MVA R-204, OAL
R-246 and MOD R-064 to MOD VOR/DME. Thence . . . .
MUSTANG TRANSITION (FMG.MOD2): From over FMG VORTAC via FMG R-182
and MOD R-064 to MOD VOR/DME. Thence . . . .
. . . . From over MOD VOR/DME via MOD R-245 to CEDES INT, then via ECA R-215
to FAITH INT/DME.

MODESTO TWO ARRIVAL
(MOD.MOD2) 95201

SAN FRANCISCO, CALIFORNIA
SAN FRANCISCO INTL

P36

SW-2, 4 NOV 1999

Illustration 3. San Francisco Airport Modesto Two Arrival

The 3-to-1 Rule

The 3-to-1 rule means that you take the altitude (in 1,000s of feet) you need to lose and multiply it by 3. This means we plan to fly 3 NM for every 1,000 feet of altitude lost. That's the distance required for most turbojet enroute descents at idle power. To use this method, an aircraft should maintain a constant mach number and/or constant indicated airspeed to stay on the enroute descent profile.

Q *You are cruising at FL230 and have been cleared to descend, pilot's discretion, to 11,000' MSL by 15 DME before the next VOR. How far out would you start your enroute descent?*

A *Start the enroute descent at 51 DME prior to the VOR.*

Step one: Figure out how much altitude there is to lose. In this case, it's 12,000 feet.

Step two: Multiply the altitude (in 1,000s) to lose by 3, which is (12 × 3) 36 NM.

Step three: Compute the end of descent point, which is 15 DME. Add the enroute descent distance, which we just figured to be 36 NM, for a total of (15 + 36) 51 DME before the next VOR to begin a normal enroute descent.

If your aircraft uses a different formula to compute enroute descents, such as 2 or 2.5 times the altitude (in 1,000s), use the factor for your aircraft rather than the 3-to-1 rule. The steps in the solution remain the same.

Also consider the extra distance needed for a slowdown to comply with a crossing restriction. In the above example, if the cruise speed at FL230 was

300 KIAS and the crossing restriction at 11,000 feet MSL included a slowdown to 250 KIAS, you would need to include an extra 1 NM per 10 knots to slow down, for a total extra distance in this case of 5 NM. The final answer, therefore, would now be to start the enroute descent at 56 DME.

An alternate way to calculate enroute descents using the 3-to-1 rule is based on dividing the flight level (altitude in hundreds of feet) by 3. This would result in a descent gradient of 300 feet per nautical mile. The earlier method of multiplying the altitude (in 1,000s) by a 3 results in a descent gradient of 333 feet per nautical mile. In my experience, the majority of pilots prefer the first method discussed of multiplying by 3. However, since enroute descent calculations are simply an approximation tool, either method is acceptable. If you decide to divide the flight level by 3, change step two from above to read "divide the flight levels, or altitude in hundreds of feet, by 3, which is (120 ÷ 3) 40 NM." The result with this revised method is an increased calculated distance for the enroute descent.

Let's now work a problem using a real world example.

Assume you are on a flight from Kansas City to San Francisco, cruising at FL350 and 300 KIAS, and using the Coaldale transition to the Modesto Two Arrival. Referring to the San Francisco Modesto Two Arrival chart on Page 40, calculate the start of an enroute descent based upon the following clearance.

"Flyways 777, you are cleared pilot's discretion to descend to cross Cedes intersection at 11,000' MSL and 250 knots as published."

 At what DME should you plan to start your enroute descent (no winds)?

A *You will need to start the descent 45 DME prior to the Modesto VOR.*

Step one: You will need to descend 24,000 feet from FL350 to 11,000' MSL.

Step two: 24 × 3 = 72 NM to descend 24,000 feet.

Step three: Since Cedes is 32 DME past the MOD VOR, subtract 72 − 32 = 40 DME prior to the MOD VOR to start the descent.

Step four: Add 5 NM + 40 NM = 45 DME to account for the slowdown from 300 KIAS during the descent to the crossing restriction of 250 KIAS at Cedes.

Constant Descent Rate

This method is typically used by high-speed piston aircraft and turboprop airplanes that maintain a constant vertical speed during the descent and relatively constant ground speed during the descent.

Let's use this method with the first example from the last section using the 3-to-1 rule. However, we will need to include additional information to solve an enroute descent problem using a constant rate descent.

Q *You are cruising at FL230 and have been cleared to descend, pilot's discretion, to 11,000' MSL by 15 DME before the next VOR. Your ground speed during the descent will be 240 knots, and your planned descent rate is 2,000 fpm. How far out would you start your enroute descent?*

A *Start your enroute descent at 39 DME.*

Step one: Calculate the altitude to lose. In this case, it's 12,000'.

Step two: Calculate the time required to descend. (12,000' ÷ 2,000 fpm) = 6 minutes.

Step three: Calculate the distance traveled during the time needed to make the constant rate descent. 240 knots ground speed is equivalent to 4 NM per minute (remember, 60 knots = 1 NM/MIN). Thus, (4 NM per minute) × (6 minutes) = 24 NM.

Step four: Combine the above steps as follows. Add the distance from Step three to the descent restriction at 15 DME; i.e., 24 + 15 = 39 DME.

Could you solve a similar problem using a constant descent rate of 1,000 fpm?

The Pitch Attitude Solution

This method is typically used in general aviation—such as single-engine piston aircraft—although the method can be quite useful and accurate for any type of aircraft. I used this method while instructing primary instrument students, yet it seems to have gotten lost in the sophistication of modern equipment. The solution requires visualizing the descent on the aircraft pitch indicator, as I will illustrate and describe below. We will be "aiming" the nose of the aircraft on the attitude indicator much the same as one would aim a rifle at a target. The mathematical basis for the end result solution is the 60-to-1 rule that was explained in the previous chapter. To ensure proper pitch changes, the explanation will assume that the starting pitch attitude at cruise altitude is at zero degrees, and that no changes will be made to the aircraft configuration that might affect a pitch change, e.g., extending the flaps. In addition, the pilot will need to adjust the power and drag devices as necessary to maintain the desired speed for the descent.

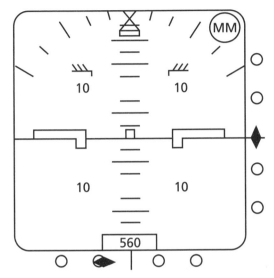

Illustration 4. Attitude Indicator

To start, let's assume that you are at FL230 at 240 knots ground speed, as in the previous example. However, this time you are given a clearance to start a descent now and to be level at 11,000' MSL in 36 NM. This method requires only two pieces of information to work the problem: altitude to lose and distance to lose that altitude. Here's how I would set up the solution for the descent using the attitude indicator.

Step one: Note the distance required to descend to the lower altitude. In this case, we need to complete the descent in 36 NM from our current position. Abeam the 10° nose-down pitch indicator line on the attitude indicator, label this line with a value of 36.

Step two: Calculate the amount of altitude to lose in thousands of feet. In this case, we need to lose 12,000 feet. Then, starting from a zero degree pitch attitude on the attitude indicator, visualize or project how far below the zero line the number 12 would be in relation to the 36 which is now at the 10° nose-low line. Since 12 is one-third of 36, then we need to project down one-third of 10°, for a 3.3° nose-low attitude. This, in fact, is the projected pitch attitude needed to comply with the descent restriction in the clearance.

This method is also handy for visual descent points during a nonprecision approach and will be discussed more in the next chapter. Here are a couple of practice problems to work. The answers are in Chapter 7.

Altitude to Lose	Distance Available	Pitch Attitude
5,000 feet	10 NM	?
23,000 feet	70 NM	?
4,000 feet	20 NM	?
7,000 feet	28 NM	?

Table 3-1

Wind Corrections During Descent

I am only going to spend a short time with calculating adjustments to the enroute descents due to headwinds or tailwinds. For most situations, this adjustment does not significantly alter the computations we have just dealt with.

If you have in your weather forecast a significant headwind or tailwind during the descent, it is reasonable to make an adjustment for the amount of time that you will be exposed to this wind. You should be able to calculate the distance that this wind would shift your enroute descent start point.

Step one: Determine how much average headwind or tailwind component you expect during your descent to the crossing restriction on your arrival procedure. Note that the forecast winds generated for you during your flight planning are provided in degrees **true north**, not magnetic north. Obviously, you will need to convert to magnetic north to be useful in the calculation. Just a rough estimate through the descent altitudes at your point of descent is adequate. I would not take the time to whip out the calculator to average both the direction and strength of the wind. Keep it simple!

Step two: Determine how much time you will use during the descent without an adjustment for wind. Even an estimate or approximation of the time is sufficient. From the higher cruise flight levels, this may be in a range from 10 to 15 minutes.

Step three: Use the time-speed-distance techniques discussed in Chapter 2 to calculate the distance adjusted for the wind.

Q *During your enroute descent from FL350 to 11,000' MSL, you estimate that you will have an average 90-knot tailwind component. How could you adjust your top of descent to account for these significant winds?*

A *12 NM earlier.*

Step one: The problem has already provided you with one part of the solution: you have a 90-knot tailwind component during the descent.

Step two: For a descent of 24,000', we will estimate a descent time of 8 minutes (assuming an average rate of descent of 3,000 fpm).

Step three: The adjusted distance equals the distance traveled in 8 minutes at a speed of 90 knots. Or, 8 minutes at 1.5 NM per minute, which equals 12 NM. In this case, since it is a tailwind, we would start down 12 NM earlier than was originally computed.

Chapter 4

Calculating Visual Descent Points

Before getting into the mathematics of calculating our own visual descent point for a nonprecision approach, I believe it would be helpful to provide some background information from the AIM and the regulations on what the Visual Descent Point (VDP) is intended to accomplish. Then, we will discuss three techniques for constructing a descent point where no VDP is published on a nonprecision approach. To begin with, here's what the AIM says:

Visual Descent Points (VDPs) are being incorporated in nonprecision approach procedures. The VDP is a defined point on the final approach course of a nonprecision straight-in approach procedure from which normal descent from the MDA to the runway touchdown point may be commenced, provided visual reference required by FAR Section

91.175(c)(3) is established. The VDP will normally be identified by DME on VOR and LOC procedures and by along track distance to the next waypoint for RNAV procedures. The VDP is identified on the profile view of the approach chart by the symbol: V.

VDPs are intended to provide additional guidance where they are implemented. No special technique is required to fly a procedure with a VDP. The pilot should not descend below the MDA prior to reaching the VDP and acquiring the necessary visual reference.

Pilots not equipped to receive the VDP should fly the approach procedure as though no VDP had been provided.

Here are a few more points from the *United States Standard for Terminal Instrument Procedures* (TERPs) that might help you understand the usefulness of a VDP:

- A VDP will be for a normal descent to touchdown, usually a 3 degree glidepath;
- If a VASI is available on the runway, the VDP will align with the VASI glidepath;
- If a VASI is not available on the runway, the VDP will provide a normal glidepath to the runway threshold.

Please note that the AIM description of VDPs states that procedurally you **should not** descend below the MDA prior to the VDP, etc. However, in FAR §121.651(c)(4), there is implied approval to descend prior to the VDP if the descent from the MDA "to the runway cannot be made using normal procedures or rates of descent if descent is delayed until reaching that point."

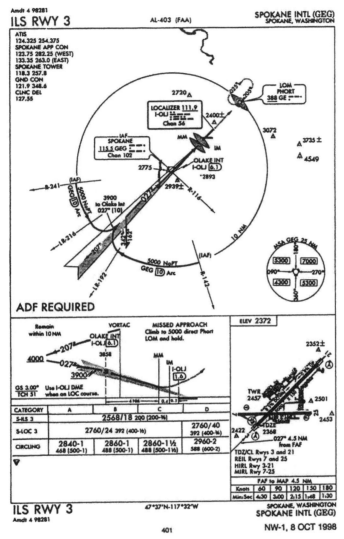

Illustration 5. ILS RWY 3 Spokane International

Every transport category aircraft I can think of should be able to comply with the intent of the VDP and not require an early descent even if we have the runway in sight. Besides, there may also be limiting obstacles to contend with, and a shallow glidepath or "dragged-in" approach can sometimes lead to visual illusions during the transition to landing.

Now that we've established some guidelines for using a VDP, what can you do in lieu of a VDP if there is not one published on the approach plate?

Some airline pilots are now using VDP-style techniques to build their own Planned Descent Point (PDP). This is simply a **tool** that enhances the normal transition to landing from the MDA, just like the VDP. Unlike the VDP, which has regulatory criteria to maintain from the TERPs handbook, a PDP is more like a pilot aid to use in the cockpit to help get the job done right. And, a PDP will never be published or have the regulatory backing of a VDP. However, it can really be a smart technique when a VDP is not otherwise available.

Here's the text version of a nonprecision approach plate that does not have a published VDP. Your goal is to calculate your own PDP. By the way, these techniques of calculating a PDP work for any type of aircraft, regardless of speed flown on the approach.

Q   *The KGEG airport ILS Rwy 3 (see Page 51) glideslope is NOTAMed out of service. Thus, you will need to fly the LOC (GS out) with a MDA of 2,760' (HAT of 392'). There is a colocated DME with the ILS frequency. The FAF is identified by D6.1. The missed approach point is at the runway threshold identified by D1.6. In addition, with a ground speed of 120 knots, the timing on the approach is 2:15. If we assume that the ceiling and visibility is adequate to plan on seeing the runway environment from the MDA, what is your calculated PDP for the approach?*

A *There are three methods to calculate a PDP, two of which are familiar to most pilots and one which is rarely used. The most familiar method involves using DME. The other two methods involve using timing on the approach or visualizing the descent on the attitude indicator as discussed in the prior chapter. All methods work well for planning purposes. I'll summarize the differences in the methods after we solve the problem. Let's start with the DME method.*

DME Method

Step one: Divide the HAT by 300. That would give us the distance in nautical miles that it would require to descend from the MDA to touchdown at the runway threshold with a 3° glidepath, or 300 feet per nautical mile. In our problem, 392 ÷ 300 = 1.31 NM. However, let's keep it simple and use 390 ÷ 300 = 1.3 NM.

Step two: Determine the DME at the runway threshold. In our problem, the runway threshold is the same point as the missed approach point at 1.6 DME.

Note: In those cases where the missed approach point is not identified by DME, yet it visually appears on the chart to be aligned with the runway threshold, and DME is used to identify the FAF, you should refer to the timing box for the approach plate to note the distance from the named FAF to the MAP. Then, subtract that distance in nautical miles from the FAF DME to get the DME at the threshold. This is always a good technique to use as a backup for your calculations because the small print on the approach plates does not always clearly portray distances on final approach.

Step three: Add the two distances calculated from Steps one and two above. This will give you an accurate DME for a PDP with a 3° glidepath to the runway threshold. In our problem, the PDP would be calculated to be at 2.91 DME, but our estimate of 2.9 DME is close enough for us to use.

Be extra cautious using this method when the DME to be used is from a VOR that is crossed earlier on the final approach; i.e., the DME may actually be getting larger the closer you get to touchdown. It may be helpful to sketch a layout of the runway versus DME source location to help clarify the math.

In fact, the Spokane ILS Rwy 3 portrays this situation. Notice that the GEG VOR is between the FAF and MAP.

Timing Method

Step one: Divide the HAT by 10. This gives us the time in seconds required to descend from the MDA to touchdown on the runway at 600 feet per minute rate of descent. In our problem, 392 ÷ 10 = 39.2 seconds, but 39 seconds is close enough.

Do you understand why we divide by 10 in the first step for the timing method? Most pilots concede that the reason is simply because that's what has always worked. True statement. However, there is a fundamental and easy reason for using 10. **Dividing the HAT by 10 is really dividing the HAT by a descent rate of 10 feet per second, or the equivalent of 600 feet per minute!** Therefore, if you choose to use this method to determine your visual descent point, the calculation forces you to make a constant 600 feet per minute rate of descent during the visual transition to landing from the MDA. That descent rate works well enough for most aircraft, but, for some higher speed aircraft 600 fpm may bring them in a little shallow on the visual glidepath.

Step two: Determine the timing required on the approach from the FAF to the MAP as shown in the timing box. It works for any chosen ground speed, but for our problem today at 120 knots ground speed, the timing, as stated, is 2:15.

Step three: Subtract the calculation in Step one from the timing in Step two. This gives us the timing from the FAF to the PDP. We always have the clock running from the time we cross the FAF anyway, so this requires no additional timing. In our problem, the timing for the PDP is at 2:15 – :39 = 1:36 on the approach.

How would the timing to the visual descent point be different if we were flying at 160 knots ground speed? In Step two, we would interpolate between 150 and 180 knots on the published timing box to calculate 1:42 as the timing for the approach at 160 knots from the FAF to the MAP. Then, in Step three, we perform the same steps as before by subtracting this new timing at 160 knots minus the same time to descend at 600 fpm from Step one. Thus, 1:42 – :39 = 1:03 is the new timing at 160 knots to the PDP. Note, however, that the glidepath at 160 knots is a little shallower than the glidepath for 120 knots.

So, what's the difference between these first two methods? The DME method uses 300 feet per nautical mile to approximate a 3° glidepath to calculate a PDP. The timing method uses a constant 600 feet per minute rate of descent as the basis for calculating the PDP timing. In essence, the two methods may define a different point in space.

However, at approximately 120 knots ground speed, the two different PDPs would be very close to each other.

Please bear in mind, also, that the formulas I've used here use some numbers that have been rounded off for ease of use. This also helps you do the work without a calculator.

Pitch Attitude Solution

This method was discussed in detail in Chapter 3 as an option to calculate enroute descents. This time, however, the same techniques for visualizing the descent on the attitude indicator in the airplane will be used to ensure an adequate rate of descent on final approach, during a nonprecision approach procedure, to see the runway environment and be in a safe position on glidepath to make the visual transition to land.

In the problem previously described for the LOC (GS out) Rwy 3 at KGEG, the distance from the FAF to the MAP was 4.5 NM. In addition, the HAT at the FAF altitude of 3,900' MSL would be 1,532' AGL (3,900' MSL minus the touchdown zone elevation of 2,368' MSL equals 1,532' AGL). The 4.5 NM and 1,532' AGL now become important to us to help determine a minimum pitch necessary to descend from the FAF for a safe landing.

Step one: Just as in the example from Chapter 3, note the distance required to descend to the lower altitude. In this case, we need to descend from the FAF altitude to the threshold in 4.5 NM. Therefore, abeam the 10° nose-down pitch indicator line on the attitude indicator, label this line with a value of 4.5 NM.

Step two: Calculate the amount of altitude to lose in thousands of feet. In this case, we need to lose 1,532' from the FAF altitude of 3,900' MSL to the threshold. Then, starting from a zero degree pitch attitude on the attitude indicator, visualize or project how far below the zero line 1.5 would be in relation to the 4.5 which is now at the 10° nose-low line. Since 1.5 is one-third of 4.5, then we need to project down one-third of 10°, for a 3.3° nose-low attitude.

This is the minimum pitch attitude, averaged over the length of the final approach course, necessary to land from the nonprecision approach procedure. If, however, you were able to use more than the minimum of 3.3°, such as a 4° or 5° nose-low attitude, during the descent from the FAF to the MDA, you should normally have the time to level at the MDA long enough to visually acquire the necessary cues to complete a normal transition to landing.

The significant difference between this pitch attitude solution versus the DME method or timing method is that it is mainly useful in determining only the minimum flight path angle (or pitch attitude) required to ensure a successful landing. Although it is possible to compute a descent point from the MDA using this method, the mental exhaustion that would result is not worth the time or effort. The goal of this study guide is to avoid those extensive mind-bending computations. Therefore, this method may not be operationally practical for airline operations and may only provide a practical teaching opportunity on final approach for lower speed general aviation aircraft. However, this technique may be useful for some approaches using VNAV (Vertical Navigation) or constant descent angle procedures during approved operations.

Visual Glidepath

In this chapter the discussions have been centered on computing visual descent points for use in the weather with nonprecision approaches. However, many times you are in clear weather, or at least VMC, and cleared the visual straight in approach to a runway. Therefore, as a reminder, be sure to also use these same techniques to compute visual glidepath checkpoints as you descend on final during a visual approach. Because it is often difficult to see the VASI or PAPI much more than three miles from the runway, a predetermined altitude at a certain distance from the threshold ensures a more stabilized approach.

The most common method is the DME method for assisting with the visual glidepath. Since you still want to use a 3° glidepath on final, you can divide the actual altitude above the airport by 300 feet per nautical mile to compute the distance from the runway you should be. Or, conversely, you can multiply your actual distance from the runway by 300 feet per nautical mile to compute the altitude above the runway that you should be at that point. Using the actual distance and computed altitude for a 3° glidepath more readily provides feedback on where the glidepath is located. Or, more practically, it indicates if you are above or below the desired glidepath.

It is common to also use a factor of 333 feet per NM for the descent gradient. This calculation is acceptable, although it adds a minor level of complexity to the mental math.

Note: Remember that a 3° glidepath is more accurately defined as a gradient of 318 feet per NM. Use of either the 300 or 333 feet per NM simply brackets the true value. As such, the two methods will compute a glidepath slightly low or slightly high on a 3° glidepath.

Here's a short table to illustrate the use of both gradients in determining a guide for a proper visual approach glidepath. Don't forget to also use the VASI or PAPI guidance once it is visible.

300'/NM	Distance	333'/NM
1500' AGL	5 NM	1666' AGL
1350' AGL	**4.5 NM**	**1500' AGL**
1200' AGL	4 NM	1333' AGL
1050' AGL	3.5 NM	1166' AGL
900' AGL	**3 NM**	**1000' AGL**
750' AGL	2.5 NM	833' AGL
600' AGL	2 NM	666' AGL
450' AGL	**1.5 NM**	**500' AGL**
300' AGL	1 NM	333' AGL

Table 4-1

The method illustrated in the right-hand column is more complicated. Three of the checkpoints are highlighted at 4.5, 3, and 1.5 NM which have simpler altitude numbers to remember of 1,500, 1,000, and 500 feet AGL.

Chapter 5

Mental Math Test

1. List the reciprocal headings for 042°, 077°, 168°, 243°, 265°, 331°.

2. Calculate the speeds that an aircraft will start hydroplaning based on tire pressures of 40, 60, 100, and 210 psi.

3. The temperature/dewpoint at an airport is 66°/60°F. Calculate the equivalent centigrade temperatures and estimate the bases of the forecast cloud layer.

4. At South Lake Tahoe airport, with an airport elevation of 6,262' MSL, the altimeter setting is 29.73 in. Hg. (QNH). What is the pressure altitude?

5. At FL220, the outside air temperature indicates -30°F. What is the temperature deviation from ISA?

6. You are descending from FL280 for landing and forgot to reset your altimeter to 30.11 in. Hg. (QNH). What will your altimeter read after landing at the airport?

7. For landing on runway 34, the tower reported winds are 030/20. What is the crosswind component and estimated drift angle with an approach true airspeed of 120 knots?

8. As your aircraft crosses directly over the LAX VOR at FL180, what DME will you indicate?

9. For a daytime VFR flight, how many gallons of Avgas are required for a flight time of 3 + 20? Your C-182 fuel burn is approximately 10 gph at 8,000' MSL cruise altitude.

10. For a nighttime IFR flight, how many gallons of Jet A is required for a flight time of 4 + 10? The time to the alternate is 0 + 25. Fuel flow in the Learjet is 900 pph at FL330.

11. After takeoff in your B-727, you have an engine failure and must return for landing. You now have 45,000 pounds of fuel on board, but can only have 17,500 pounds of fuel to not exceed maximum landing weight. At a fuel dump rate of 2,500 ppm, how long will it take you to dump fuel for landing?

12. How much total fuel will you dump in 13 minutes at a rate of 1,100 ppm?

13. Using your magnetic compass only for heading reference, starting at 110 KIAS, 5,000' MSL, heading 030°, and latitude 35° North, determine the SRT bank angle and lead point for a right turn to heading 190°.

14. On the VOR 23L approach plate for ABC airport, there is a 20 DME arc transition from the ABC 355/20 (Yellow) to the ABC 052/20 (Green). What is the distance along the arc from Yellow to Green?

15. When flying the same VOR 23L transition, how many degrees "lead" would you need to start the turn from the arc to the final approach course of 232°, using a standard rate turn at 180 knots?

16. With a 225 KTAS and 45 knot tailwind, how far will you travel in 12 minutes?

17. If you travel 300 NM in 90 minutes, what is your ground speed?

18. You are cruising at FL220 and 280 KIAS and have been cleared to descend, pilot's discretion, to 8,000' MSL by 8 DME past the next VOR. How far out would you start your descent?

19. You are cleared to start an immediate descent from FL190 to 12,000' MSL. You have 35 DME remaining to complete the descent. What is your estimated pitch attitude for the descent?

20. You have been cleared the visual approach for landing at Bakersfield (airport elevation of 507' MSL). For a normal glidepath, what should your altitude be at 6 NM from the threshold of the runway?

21. The KMRT airport ILS Rwy 4 glideslope is NOTAMed out of service. Thus, you will need to fly the LOC (GS out) with a MDA of 1,760' (427'). There is a colocated DME with the ILS frequency. The FAF is identified by D5.9. The missed approach point is at the runway threshold identified by D1.4.

In addition, with a ground speed of 150 knots, the timing on the approach is 1:48. If we assume that the ceiling and visibility is adequate to plan on seeing the runway environment from the MDA, what is your calculated PDP for the approach?

Chapter 6

Summary

Did you realize before completing this study guide that there were so many math calculations that you needed to be doing in the cockpit? It is mind-stretching to keep up with all of this during the progress of just one flight. Yet, after the realization that all of these maneuvers and events really do occur, a conscious effort to practice the various computations will result in an obvious increase in your skill and professionalism.

As a reminder, once you have discovered that you need to calculate a math solution to any problem, first define the problem; i.e., what is the answer I'm looking for? Second, look for the right formula to use. Most of the formulas you will ever need are right here in this book. Third, rearrange the formula to solve for the answer that you need. And, fourth, plug in the numbers and solve.

Since our goal here is to do these problems without the assistance of pen, paper, or calculator, I recommend that you practice the problems over and over. First, do the problems on paper. Second, study and memorize the steps of the problems and the relationship between the variables of the problem. Third, after practice on paper, push the paperwork aside and visualize the exact same steps without writing them down. Once you can do this repeatedly, you are well on your way to proficiency in cockpit mental math skills.

Don't be afraid to modify some of the formulas and methods I've explained here. Especially for the flight maneuvers and profiles that might differ for your airplane or company. There are as many techniques and formulas to safely and professionally fly airplanes as there are skilled pilots and types of aircraft. Use caution, though; be sure that you do truly understand the *why* of a formula or number that has been taught to you. Without an appropriate level of understanding of these mental math skills, the results might be sloppy, unprofessional, inefficient, or worse—unsafe.

Therefore, I'd like to encourage you to stay proficient in these mental math skills, honing them with every opportunity. In fact, many times you can "beat" the computer, flight management system, or glass cockpit displays by working the problem in your head faster than the computer (especially if you have to type in the data).

We can take pride as professional pilots at any level of aviation that we encourage each other to greater levels of proficiency. Keep it up! Mentorship is a magnificent way to ensure the credibility and safety of our profession. Therefore, pass on the knowledge and skill gained through this study guide. In fact, many a veteran pilot would appreciate a new "trick" to put in his flight bag, just as a new student pilot is eager to learn from the "hangar flying" on weather days when the retired jet-jocks come down to the FBO just to see what's going on!

Good luck to you! Keep those skills sharp! I hope to see you flying the line very soon!

Chapter 7

Answer Key

Chapter 2—Airborne Math Problems

Tire Pressure	V_{HP}
50 psi	63 knots
120 psi	99 knots
150 psi	110 knots
230 psi	135 knots

Table 2-4

°C	Technique #1	Technique #2	Technique #3	°F
12°C	54°F	54°F	54°F	Calculate °F
25°C	77°F	77°F	80°F	Calculate °F
0°C	32°F	32°F	30°F	Calculate °F
Calculate °C	5°C	4°C	5°C	40°F
Calculate °C	27°C	27°C	26°C	81°F
Calculate °C	22°C	22°C	21°C	72°F

Table 2-6

Altitude	ISA Temp	Actual Temp	Temp Dev
5,000 MSL	5°C	20°C	+15°C
8,000 MSL	-1°C	15°C	+16°C
FL 210	-27°C	-10°C	+17°C
FL 350	-55°C	-60°C	-5°C

Table 2-7

Wind Angle to Runway	Total Wind Strength	Crosswind Component
030	20	10 knots
050	20	14 knots
070	18	16 knots

Table 2-9

TAS	Crosswind Component	Drift Angle
150 knots	12 knots	≈ 5°
360 knots	48 knots	= 8°
90 knots	30 knots	= 20°
0.78 Mach	50 knots	≈ 7°

Table 2-10

Given:	Find:
200 Knots	230 MPH
180 MPH	156 Knots
8 MPS	16 Knots
9 KM	6 SM

Table 2-13

Gallons	Pounds
55 gal. Avgas	330 lbs. Avgas
80 gal. Avgas	480 lbs. Avgas
167 gal. Avgas	1000 lbs. Avgas
500 gal. Jet A	3350 lbs. Avgas
750 gal. Jet A	5000 lbs. Jet A
1275 gal. Jet A	8500 lbs. Jet A

Table 2-16

Dump Rate	Time	Fuel Dumped
1,300 PPM	5 min.	6,500 lbs
2,500 PPM	18 min.	45,000 lbs
3,000 PPM	6 min. 20 sec.	19,000 lbs
2,500 PPM	12 min.	30,000 lbs
2,200 PPM	5 min.	11,000 lbs
1,500 PPM	7 min.	10,500 lbs
1,200 PPM	11 min.	13,200 lbs
2,500 PPM	5 min.	12,500 lbs
3,000 PPM	16 min.	48,000 lbs
2,000 PPM	10 min.	20,000 lbs

Table 2-17

Bank Angle (Left/ Right)	Start Heading	Desired Heading	Latitude	Lead Point
15° R	270°	360°	45° North	310°
15° L	270°	180°	34° North	151°
25° R	360°	090°	40° North	082°
20° R	090°	190°	40° North	223°

Table 2-19

DME Arc	Lead Radials
15 NM	4 degrees
20 NM	3 degrees
12 NM	5 degrees

Table 2-21

KTAS	Standard Rate Turn Bank Angle	Turn Radius (SRT/30° Bank Max)
90 KTAS	13.5°	0.5 NM
120 KTAS	18°	0.6 NM
200 KTAS	30°	1 NM
0.80 Mach	30° (max IFR)	6 NM

Table 2-22

KIAS	Altitude	KTAS
100 KIAS	10,000 ft MSL	120 KTAS
140 KIAS	5,000 ft MSL	154 KTAS
200 KIAS	13,000 ft MSL	252 KTAS
280 KIAS	FL350	476 KTAS

Table 2-23

KTAS	Wind	Time	Distance
240	60 TW	40 min.	200 NM
280	70 HW	10 min.	35 NM
150	0	2 min.	5 NM
300	0	4 min.	20 NM
420	60 TW	50 min.	400 NM
420	0	2 min.	14 NM
400	0	1.5 hr.	600 NM
500	0	45 min.	375 NM
510	0	40 min.	340 NM

Table 2-25

Chapter 3—Calculating Enroute Descents

Altitude to Lose	Distance Available	Pitch Attitude
5,000 feet	10 NM	5.0° down
23,000 feet	70 NM	3.3° down
4,000 feet	20 NM	2.0° down
7,000 feet	28 NM	2.5° down

Table 3-1

Chapter 5—Mental Math Test

1. 222°, 257°, 348°, 063°, 085°, 151°.

2. Approximately 57 knots, 70 knots, 90 knots, and 130 knots.

3. Temperature/dewpoint is 19°/16°C. The forecast cloud base is estimated at 1,333' AGL using 6°F ÷ 4.5°F/1,000' = 1,333' AGL; or, 1,200' AGL using 3°C ÷ 2.5°C/1,000' = 1,200' AGL.

4. 6,262' MSL elevation plus 190' altimeter correction equals 6,452' pressure altitude.

5. FL220 ISA equals -29°C (-20°F). Therefore, temperature deviation equals -10°F or -5°C.

6. The altimeter will read 190' low at the airport. (30.11 − 29.92 = .19 in. Hg. which is equivalent to 190' altitude).

7. 14 knots of crosswind with a drift angle of 7 degrees.

8. 3 DME.

9. 38.3 gallons Avgas. Don't forget the 0 + 30 reserve requirement.

10. 720 gallons Jet A. Don't forget the 0 + 45 reserve requirement.

11. 11 minutes.

12. 14,300 pounds.

13. At 110 KIAS, 5,000' MSL, TAS = 121 knots. SRT bank angle = 18°. When turning to south, plan on overshoot minus lead, or, 190° + 35° (latitude) − 6° ($\frac{1}{3}$ of 18°) = 219° lead point on magnetic compass.

14. 57 degrees of arc ÷ 3°/NM = 19 NM.

15. Just under 3° lead radial required for turn to final course.

16. 54 NM.

17. 200 knots.

18. 37 DME prior to the VOR.

19. 2° nose-down pitch.

20. Either 1,800' AGL (6 NM ´ 300'/NM) + 507' MSL (field elevation) = 2,307' MSL at 6 NM from runway, or, 2,000' AGL (6 NM ÷ 333'/NM) + 507' MSL (field elevation) = 2,507' MSL at 6 NM from runway.

21. PDP = 2.8 DME or 1:05 timing from the FAF.

Appendix A: Basic Math Exercises

14 + 13 = 27	9 + 26 = 35	27 + 27 = 54	39 + 64 = 103
58 + 79 = 137	8 + 35 = 43	27 + 32 = 59	175 + 180 = 355
121 + 200 = 321	180 + 75 = 255	23 – 12 = 11	37 – 28 = 9
66 – 33 = 33	113 – 32 = 81	144 – 76 = 68	59 – 43 = 16
61 – 43 = 18	61 – 53 = 8	317 – 24 = 293	347 – 180 = 167

Table A-3

12 x 3 = 36	15 x 9 = 135	13 x 6 = 78
28 x 3 = 84	15 x 7 = 105	13 x 12 = 156
27 ÷ 9 = 3	48 ÷ 30 = 1.6	14000 ÷ 5 = 2800
81 ÷ 9 = 9	960 ÷ 30 = 32	18 ÷ 7 = 2.6

Table A-6

$7^2 = 49$	$9^2 = 81$	$13^2 = 169$
$\sqrt{7} \approx 2.6$	$\sqrt{9} = 3$	$\sqrt{150} \approx 12$

Table A-8

Appendix A

Basic Math Exercises

Addition and Subtraction

First, I will show you how one skill leads to another by going over some very basic addition problems. As a result, your overall mental math skill level will rise significantly with the correct and continuing application of basic addition and subtraction skills. As we examine and practice problems remember that the same rules or discipline required to add single-digit numbers also apply to larger numbers. In other words, don't allow large numbers to overwhelm you. We will apply the same skills we used with small numbers—we just have to repeat the steps a couple of more times per problem! Mastery of addition skills contributes directly to the mastery of subtraction skills and vice versa.

As we proceed through the demonstration problems and explanations that follow I want you to look for the common features, relationships, differences, and techniques used to solve the various problems. After we develop a foundation of basic skills, we will continue to build your skills using those relationships and techniques to increase your proficiency. Keep in mind that the intent is to master skills instrumental to **mental** math problem solving. In other words, you may not have pen, paper and/or calculator handy to help you out, so I will keep my explanations as simple and clear as possible to help you solve problems in the cockpit. You can refer to Appendix B to review basic math tables for addition and subtraction.

Demonstration problems:

3	6	12	48	67
+ 4	+7	+24	+34	+49
7	13	36	82	116

Table A-1

Each of these addition problems builds upon the skill of the prior problems. In the first problem, the addition of 3 + 4 is relatively simple and straightforward. The second problem shares the simplicity, but requires a two-digit answer; e.g., 6 + 7 = 13. Both require memorizing the basic addition table in Appendix B.

Problem three goes to the next level of adding two two-digit numbers. In this case, add the numbers from the right column (ones digit), then add the numbers from the left column (tens digit). From the right column, 2 + 4 = 6. From the left column, 1 + 2 = 3 (or 30 since the 3 is in the tens column). Joining the two steps results in 36 as the combined answer (30 + 6).

Problem four again raises the level of skill by building on the first three approaches. First, add the ones column. 8 + 4 = 12. Since the answer in the ones column can only use a single digit from the ones column in the answer, we keep the 2 in the ones column in the answer. From this first step, use the additional 10 as a carry-over of 1 to the tens column. Now, we have 4 + 3 + 1 = 8 in the tens column. Joining the two steps results in 82 as the combined answer (80 + 2).

At this point, you should start to see how you go from a basic level to more complex levels by using the common features and building block approach to solving problems. Let's continue with one more level of problem solving. Then I will offer a useful technique for mental math problem solving.

Problem five adds 67 + 49 = 116. To solve, add the ones column, then the tens column plus any carry-over from the ones column. Thus, 7 + 9 = 16, or 6 in the ones column plus a 1 to carry-over to the next tens column. Then 6 + 4 + 1 in the tens column equals 11, or 110, since 11 is in the tens column. Hence, 6 + 110 = 116.

One useful technique in addition and subtraction problems is to initially "round-off" the numbers to make them easier to work with. It is much easier to add numbers that don't have carry-over; that is, numbers that are in even tens. For example, in the last problem we added 67 + 49. It is easier to first add the closest even tens numbers, followed by correcting for the rounding. Thus, the first step would be to add 70 + 50, which equals 120. Then, correct the differences and adjust the final answer. Hence, the difference from 70 to 67 would be -3, and the difference from 50 to 49 would be -1, for a total difference of -4 from the total. Thus, 120 − 4 = 116.

Obviously enough, this is the original correct answer, but perhaps seeing this relationship allows you to solve it more easily in your head.

Subtraction is numerically the opposite of addition. However, it seems a little more difficult to perform. Nonetheless, if you maintain the mindset that subtraction is merely working an addition problem in reverse, you are well on your way to mastering the discipline.

Demonstration problems:

7	31	46	53	112
− 3	− 4	− 19	− 19	− 58
4	27	31	34	54

Table A-2

In the first problem, $7 - 3 = 4$, is the most basic of subtractions. You must master this level first. A subtraction table for review is included in Appendix B.

In the second problem, $31 - 4 = 27$, draws upon a slightly higher level of subtraction. In a similar fashion to addition problems, first subtract the ones column on the right-hand side, then the tens column on the left. In this case, since $1 - 4$ is less than zero, we can "borrow" a 1 (which is really 10) from the tens column, with a remainder of 2 (which is really 20) in the tens column. Add the "borrowed" 10 to the 1 in the ones column, and now have $11 - 4$. Finally, with a 2 remaining in the tens column, joining the steps now leaves $20 + 7$ as the answer, or 27.

The third problem combines the methods from the prior two. First, subtract the ones column, then subtract the tens column. In this example, there is no "borrowing" to consider. Hence, $6 - 5 = 1$ in the ones column; $4 - 1 = 3$ in the tens column (or 30). Joining the two steps equals $30 + 1$, or 31.

In problem four, use a combination of the prior three methods. In the first step of subtracting the ones column, 3 – 9, you will need to "borrow" a 1 from the tens column (which is really 10). Now you have 13 – 9 = 4 in the ones column. In the tens column, you are left with a 4 (since you had borrowed 1) – 1 = 3, or 30. Combining the steps leaves 30 + 4, or 34.

The last problem simply uses larger numbers. 112 – 58 = 54. First, in the ones column, 2 – 8 becomes 12 – 8 (after "borrowing" from the tens column) which equals 4. Second, in the tens column, 0 (remember we borrowed a 1) – 5 becomes 10 – 5 (after borrowing a 1 (or 100) from the hundreds column) which equals 5. Hence, the combined answer is 54.

You can also use the rounding technique I described before with addition problems. That is, it is much simpler to subtract numbers that do not involve borrowing or carry-over between columns. In this last problem of 112 – 58, round off the numbers to 110 – 60 which equals 50. Then adjust the answer by remembering that you started off with 2 more than 110, so now you would have 2 more in the answer, or 52. Plus, you would correct for 2 less being subtracted which results in 2 more in the adjusted answer, which is now a total of 54.

This technique is no substitute for having to do the math eventually, but it is easier to start off in simpler or rounded numbers. There are many plausible techniques for quickly adding and subtracting numbers; but, since the goal here is to solve the problems mentally and without the use of pen, paper, calculator, or abacus, I suggest we concentrate on simple and straightforward techniques. Here are a few practice problems. The answers are in Chapter 7.

Practice Problems—Addition and Subtraction

14 + 13 =	9 + 26 =	27 + 27 =	39 + 64 =
58 + 79 =	8 + 35 =	27 + 32 =	175 + 180 =
121 + 200 =	180 + 75 =	23 – 12 =	37 – 28 =
66 – 33 =	113 – 32 =	144 – 76 =	59 – 43 =
61 – 43 =	61 – 53 =	317 – 24 =	347 – 180 =

Table A-3

Multiplication and Division

Let's start with a basic table of multiplication, shown in Appendix B. With practice, you should be able to recite this multiplication table from memory and without hesitation. If you haven't already memorized much of this, get started now. This is an important part of solving mental math problems. Especially with the larger numbers, repetition is the key to proficiency.

Demonstration problems:

5	9	8	15	15
x 7	x 6	x 7	x 8	x 80
35	54	56	120	1200

Table A-4

The first three demonstration problems are straight from the multiplication table in Appendix B, Table 3. As I mentioned before, you should be able to recite that multiplication table by memory. I have found, however, that even rote memory does not solve the problem when we are distracted or busy in the cockpit. Aside from practice, practice, and more practice, I will suggest a technique for simplifying multiplication mental math.

First, study the relationships of the multiplier columns in the Appendix B Table 3 Multiplication Tables. If you examine one column at a time (let's use the × 3 column), do you see a relationship between the answers as you proceed down from 1 × 3 through 1 × 12? This critical relationship is that each successive answer differs by a value of 3 from the previous or next answer. Thus, if you are able to remember that 10 × 3 = 30, you might deduce that 9 × 3 = 3 less than the prior equation; thus, 9 × 3 = 30 – 3, or 27. You can directly apply this relationship throughout the multiplication table. Using the same × 3 column, you could similarly deduce that 12 × 3 has the same answer as 10 × 3 = 30, plus 2 × 3 = 6; i.e., 36. Now try the same technique using the × 7 column with some problems you construct.

The fourth problem, 15 × 8 = 120, also requires a two-step process to arrive at the answer. First, separate the 15 into two separate columns consisting of a 5 and a 10. Second, multiply each column separately; e.g., 5 × 8 = 40, then 10 × 8 = 80, for a total of 120. Thus, the answer is the addition of the two separate multiplication problems, or 40 + 80 = 120.

The fifth problem, 15 × 80 = 1,200, is an extension of the previous problem. The difference is a factor of 10, or, rather than 8 from the prior problem it is now 10 × 8 or 80. Thus, the answer is also larger by a factor of 10. Therefore, rather than 120, the answer to 15 × 80 is now 120 × 10 or 1,200. Proficiency multiplication of large numbers comes in handy for calculating enroute descents, fuel dumping, and time-speed-distance problems.

Division is numerically the opposite of multiplication, just as subtraction is numerically the opposite of addition. In Appendix B, Table 4 and Table 6 will be very helpful in recognizing mathematical intervals and relationships between numbers in a division problem. In both tables, I have included both the fractional and decimal equivalent. As I have already emphasized, proficiency and an understanding of the fundamental patterns in division can only be successfully achieved through repetition.

From Appendix B, Table 4, study the column with the heading "÷ 6." As you proceed from the top to the bottom of the column, do you notice the interval or difference between each successive answer? Each interval is a difference of 0.16 or 0.17 (depending on how the third decimal place was rounded off). Thus, if you can remember that $9 \div 6 = 1.50$, then you can readily deduce that $8 \div 6 = 1.50 - 0.17$, or 1.33. Or, that $10 \div 6 = 1.50 + 0.17$, or 1.67. This pattern of relationship exists throughout the division table in Appendix B; that is, you can apply the same method or pattern to other division operations.

Demonstration problems:

$6 \div 2 = 3$	$9 \div 5 = 1.8$	$10 \div 3 = 3.33$	$40 \div 60 = 0.67$
$60 \div 20 = 3$	$18 \div 10 = 1.8$	$1000 \div 3 = 333.3$	$450 \div 300 = 1.5$

Table A-5

With these division problems, take particular notice of the relationship between the pairs of problems. In the first problem, $6 \div 2 = 3$, the answer is the same as for the equivalent problem that is 10 times larger in both the numerator and denominator; i.e., $60 \div 20 = 3$. Observing this simple relationship is critical to maintaining a simplified approach to mental math. Do you notice the same type of relationship in the second pair, 9 ÷ 5 versus 18 ÷ 10 = 1.8? The 9 and the 5 are both multiplied by 2, which results in the 18 ÷ 10. Both have the same answer of 1.8. The same approach could be used for 36 ÷ 20 or 81 ÷ 45, which were each multiplied (numerator and denominator) by 4 and 9 respectively. The answer remains the same: 1.8. Do you understand that when both the numerator and denominator are both multiplied by the same number that it's the same as multiplying by "1"? In other words, 9 ÷ 5 is the same as $(2 \times 9) \div (2 \times 5) = (1 \times 9) \div (1 \times 5) = 1.8$, etc.

The last four demonstration problems are intended to show that large numbers can also have simple solutions. 10 ÷ 3 is a common problem to solve. From the Division Table in Appendix B, Table 4, the answer is 3.33. Knowing this relationship leads to a myriad of other problems that can be readily solved. If the problem was 100 ÷ 3 or 1,000 ÷ 3, or 10,000 ÷ 3, the answer is just as easily solved simply by moving the decimal point to the right for each additional zero in the numerator. The answers become 33.3, 333.3, and 3,333.3, respectively. Likewise, if the problem is now 10 ÷ 6, how could we use our prior example to help solve this one? Hint: We are dividing by a number that is twice as large as the example; therefore, the answer should be half as large.

Finally, the last two demonstration problems are meant to help you seek a relationship between the numerator and denominator. 40 ÷ 60 can be quickly reduced to 4 ÷ 6, or 0.67 from the Division Table in Appendix B. The last problem, 450 ÷ 300, shows me clearly that the first number is half again as big as the second number; thus, the answer is $1\frac{1}{2}$ or 1.5.

Practice Problems—Multiplication and Division

12 x 3 =	15 x 9 =	13 x 6 =
28 x 3 =	15 x 7 =	13 x 12 =
27 ÷ 9 =	48 ÷ 30 =	14000 ÷ 5 =
81 ÷ 9 =	960 ÷ 30 =	18 ÷ 7 =

Table A-6

Squares and Square Roots

As you have been diligently progressing through this study of addition, subtraction, multiplication, and division your skill level has increased to the point where you should be more comfortable looking at number problems and recognizing a relationship or technique to provide the answer. Now, however, for the purposes of mental math for pilots, I must recommend only one method for calculating or estimating squares or square roots: rote memorization (Appendix B, Table 5). I believe this is a reasonable approach due to the limited application and/or range of values normally encountered when computing hydroplaning speeds, lift equations, etc.

Demonstration problems:

$2^2 = 4$	$15^2 = 225$	$11^2 = 121$
$\sqrt{2} = 1.41$	$\sqrt{225} = 15$	$\sqrt{120} \approx 11$

Table A-7

These demonstration problems reflect formulas you may encounter while planning a flight. During training for your pilot ratings, you studied math equations relating to aerodynamics and performance. Many of those equations required use of squares and square roots in order to solve. Although the appearance of a squared value or square root appears to make the equation much more complex, the application of the problem solving is merely the next step beyond multiplication and division. For that reason, it is critical that you master those skills to have a better grasp of solving for squares and square roots.

The first equation, $2^2 = 4$, might be a part of the lift equation, $L = (\frac{1}{2}\rho V^2)$ n C_L S, where $V = 2$ (or going twice as fast as you started). The result is that flying 2 times as fast produces 4 times the lift due to the V^2 term (assuming no other values in the equation change). [L = lift, ρ = air density, V = velocity, n = load factor (or Gs), C_L = coefficient of lift, and S = wing surface area]

Secondly, the equation $\sqrt{2} = 1.41$ might be useful in determining the increased stall speed V when you are flying with 2 Gs (n = 2). Solving for the new stall speed results in an increase of 141% or 1.41 times the original stall speed.

The rest of the demonstration problems relate to the subject of determining hydroplaning speeds. A commonly accepted formula for determining your hydroplaning speed is $V_{HP} = 9 \sqrt{P}$, where P is the value for the tire pressure (in psi) for your aircraft. My intent is to demonstrate that for tire pressures of between 121 psi to 225 psi, the range of hydroplaning speeds would vary from $9 \times \sqrt{121}$, to $9 \times \sqrt{225}$; or, to solve the equation, a range of 99 to 135 knots.

If you are aware of the squares of 11 through 16 and the approximate range of square root values of numbers between 100 through 250, you can probably solve any hydroplaning problem. Another hint:

a difference of 30 psi tire pressure will only make a difference of about 1 in the $\sqrt{P}$ term, which is then multiplied by a factor of 9 in the equation. In other words, with an increase of 30 psi, you will increase V_{HP} by approximately 9 knots, and an increase of 60 psi will increase V_{HP} by approximately 18 knots, and so on. For review, study Appendix B Table 5, Square Roots and Squares Table.

Practice Problems—Squares and Square Roots

$7^2 =$	$9^2 =$	$13^2 =$
$\sqrt{7} \approx$	$\sqrt{9} =$	$\sqrt{150} \approx$

Table A-8

There is a review test covering practical applications of addition, subtraction, multiplication, division, squares, and square roots in Chapter 5. Answers for the Appendix A practice problems and the Chapter 5 test are in Chapter 7. As you study these methods and problems, remember that the purpose of this study guide is to gain enough proficiency to complete the problems **without outside aids** such as calculator or pen and paper. However, while you are gaining proficiency, you may initially want to work the problems with a calculator or handwritten notes. Next, study the work and process you have just written down. Then repeat the problem without taking a look at your notes; rather, try to visualize the work you have just completed and formulate the answer mentally.

Interpolation

A short review of ways to interpolate between numbers in a table is necessary to complete the goal of simplifying the mental math process. Since the goal of interpolation is to calculate an exact number that lies *between* other known values (e.g., in a row or a column in a chart or table), then the exercise for us is to simplify the process into a couple of straightforward steps that can work for all problems. The demonstration problem will ask for a solution that requires interpolation between charted values in a column *and* charted values in a row.

Let's look at an excerpt from a fictitious table from a flight manual.

Aircraft Service Ceiling (Feet MSL)

Gross Weight	ISA	ISA + 10	ISA + 20
28000 lbs.	26200	25200	24200
24000 lbs.	28500	27400	26300
20000 lbs.	30900	30500	29900

Table A-9

For this example, the aircraft currently weighs 25,000 pounds and the outside air temperature is ISA + 15 degrees. What is the aircraft service ceiling?

Step one: Identify the corners that limit the answer. In other words, the final answer will be between these values. In this case, the corners would be:

25200	?	24200
27400		26300

Table A-10

Step two: Estimate or calculate how far "into" the gross weight and temperature parameters at the answer will be found. In other words, the gross weight of 25,000 pounds is 25% of the way between the bottom and the top line answers, and the temperature is halfway between the two columns. For more complex numbers that are not easily calculated, just round the percentage to a convenient number for you to use.

Step three: Calculate the answers by using one parameter at a time. Let's do the temperature of ISA + 15 first. Since this is exactly halfway between the two columns, our new available answers are now exactly between the **25200** and **24200** and exactly between the **27400** and **26300**. That gives us answers of 24,700 on the top line and 27,850 on the bottom line.

Next, using the gross weight of 25,000 pounds, let's calculate the interpolated answer that is 25% from the bottom (**27850**) up to the top (**24700**). The total difference between these answers is 3,150 feet. How about rounding this to 3,200 feet — that will make it easier to work with and not affect the answer much at all. One fourth (25%) of 3,200 feet is 800 feet. Thus, subtract 800 feet from the bottom answer in this example (**27850**), and discover that the service ceiling is now approximately **27050**.

For my own purposes, an answer of 27,000 feet is certainly close enough!

Appendix B

Reference Tables and Charts

	+ 1	+ 2	+ 3	+ 4	+ 5	+ 6	+ 7	+ 8	+ 9
1	2	3	4	5	6	7	8	9	10
2	3	4	5	6	7	8	9	10	11
3	4	5	6	7	8	9	10	11	12
4	5	6	7	8	9	10	11	12	13
5	6	7	8	9	10	11	12	13	14
6	7	8	9	10	11	12	13	14	15
7	8	9	10	11	12	13	14	15	16
8	9	10	11	12	13	14	15	16	17
9	10	11	12	13	14	15	16	17	18

Table B-1. Addition Tables

Example: 8 + 7 = 15

	- 1	- 2	- 3	- 4	- 5	- 6	- 7	- 8	- 9
1	0	- 1	- 2	- 3	- 4	- 5	- 6	- 7	- 8
2	1	0	- 1	- 2	- 3	- 4	- 5	- 6	- 7
3	2	1	0	- 1	- 2	- 3	- 4	- 5	- 6
4	3	2	1	0	- 1	- 2	- 3	- 4	- 5
5	4	3	2	1	0	- 1	- 2	- 3	- 4
6	5	4	3	2	1	0	- 1	- 2	- 3
7	6	5	4	3	2	1	0	- 1	- 2
8	7	6	5	4	3	2	1	0	- 1
9	8	7	6	5	4	3	2	1	0

Table B-2. Subtraction Tables

Example: 6 – 3 = 3

	x 2	x 3	x 4	x 5	x 6	x 7	x 8	x 9	x 10	x 11	x 12
1	2	3	4	5	6	7	8	9	10	11	12
2	4	6	8	10	12	14	16	18	20	22	24
3	6	9	12	15	18	21	24	27	30	33	36
4	8	12	16	20	24	28	32	36	40	44	48
5	10	15	20	25	30	35	40	45	50	55	60
6	12	18	24	30	36	42	48	54	60	66	72
7	14	21	28	35	42	49	56	63	70	77	84
8	16	24	32	40	48	56	64	72	80	88	96
9	18	27	36	45	54	63	72	81	90	99	108
10	20	30	40	50	60	70	80	90	100	110	120
11	22	33	44	55	66	77	88	99	110	121	132
12	24	36	48	60	72	84	96	108	120	132	144

Table B-3. Multiplication Tables

Example: 4 x 9 = 36

	÷ 2	÷ 3	÷ 4	÷ 5	÷ 6	÷ 7	÷ 8	÷ 9	÷ 10	÷ 11	÷ 12
1	1/2 0.5	1/3 0.33	1/4 0.25	1/5 0.2	1/6 0.16	1/7 0.14	1/8 0.12	1/9 0.11	1/10 0.1	1/11 0.09	1/12 0.08
2		2/3 0.67	1/2 0.5	2/5 0.4	1/3 0.33	2/7 0.28	1/4 0.25	2/9 0.22	1/5 0.2	2/11 0.18	1/6 0.16
3	3/2 1.5		3/4 0.75	3/5 0.6	1/2 0.5	3/7 0.43	3/8 0.38	1/3 0.33	3/10 0.3	3/11 0.27	1/4 0.25
4	4/2 2	4/3 1.33		4/5 0.8	2/3 0.67	4/7 0.57	1/2 0.5	4/9 0.44	4/10 0.4	4/11 0.36	1/3 0.33
5	5/2 2.5	5/3 1.67	5/4 1.25		5/6 0.83	5/7 0.71	5/8 0.63	5/9 0.55	1/2 0.5	5/11 0.45	5/12 0.41
6	6/2 3	6/3 2	3/2 1.5	6/5 1.2		6/7 0.86	3/4 0.75	2/3 0.67	3/5 0.6	6/11 0.55	1/2 0.5
7	7/2 3.5	7/3 2.33	7/4 1.75	7/5 1.4	7/6 1.17		7/8 0.88	7/9 0.78	7/10 0.7	7/11 0.64	7/12 0.58
8	8/2 4	8/3 2.67	8/4 2	8/5 1.6	4/3 1.33	8/7 1.14		8/9 0.89	8/10 0.8	8/11 0.73	2/3 0.67
9	9/2 4.5	3/1 3	9/4 2.25	9/5 1.8	3/2 1.5	9/7 1.28	9/8 1.12		9/10 0.9	9/11 0.82	3/4 0.75
10	10/2 5	10/3 3.33	5/2 2.5	2/1 2	5/3 1.67	10/7 1.42	5/4 1.25	10/9 1.11		10/11 0.91	5/6 0.83
11	11/2 5.5	11/3 3.67	11/4 2.75	11/5 2.2	11/6 1.83	11/7 1.57	11/8 1.38	11/9 1.22	11/10 1.1		11/12 0.92
12	6/1 6	4/1 4	3/1 3	12/5 2.4	2/1 2	12/7 1.71	3/2 1.5	4/3 1.33	6/5 1.2	12/11 1.09	

Table B-4. Division Tables

Example: 9 ÷ 6 = 1.5

Note: All fractions are expressed in the lower common denominator. All decimals are expressed to the closest two decimal places.

Number Squared	Number	Square Root
1	1	1
4	2	1.41
9	3	1.73
16	4	2
25	5	2.24
36	6	2.45
49	7	2.65
64	8	2.83
81	9	3
100	10	3.16
121	11	3.32
144	12	3.46
169	13	3.61
196	14	3.74
225	15	3.87
256	16	4

Table B-5. Square Roots and Squares Table

Example: $\sqrt{14} = 3.74$

$7^2 = 49$

Fraction	Decimal
1/12	0.083
1/11	0.091
1/10	0.1
1/9	0.111
1/8	0.125
1/7	0.143
1/6	0.167
1/5	0.2
1/4	0.25
1/3	0.333
1/2	0.5

Table B-6. Table of Fractions and Decimal Equivalents

Example: 0.167 = 1/6

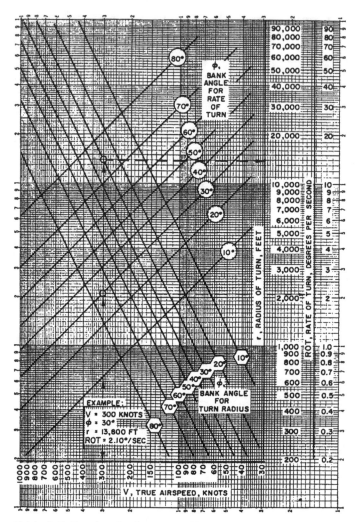

Table B-7. Turning Performance Chart

TEMPERATURE - CELSIUS / FAHRENHEIT

C°	↓	F°	C°	↓	F°	C°	↓	F°	C°	↓	F°
-60.0	-76	-104.8	-30.0	-22	-7.6	0	32	89.6	30.0	86	186.8
-59.4	-75	-103.0	-29.4	-21	-5.8	0.6	33	91.4	30.6	87	188.6
-58.9	-74	-101.2	-28.9	-20	-4.0	1.1	34	93.2	31.1	88	190.4
-58.3	-73	-99.4	-28.3	-19	-2.2	1.7	35	95.0	31.7	89	192.2
-57.8	-72	-97.6	-27.8	-18	0.4	2.2	36	96.8	32.2	90	194.0
-57.2	-71	-95.8	-27.2	-17	1.4	2.8	37	98.6	32.8	91	195.8
-56.7	-70	-94.0	-26.7	-16	3.2	3.3	38	100.4	33.3	92	197.6
-56.1	-69	-92.2	-26.1	-15	5.0	3.9	39	102.2	33.9	93	199.4
-55.6	-68	-90.4	-25.5	-14	6.8	4.4	40	104.3	34.4	94	201.2
-55.0	-67	-88.6	-25.0	-13	8.6	5.0	41	105.8	35.0	95	203.0
-54.4	-66	-86.8	-24.4	-12	10.4	5.6	42	107.6	35.6	96	204.8
-53.9	-65	-85.0	-23.9	-11	12.2	6.1	43	109.4	36.1	97	206.6
-53.3	-64	-83.2	-23.3	-10	14.0	6.7	44	111.2	36.7	98	208.4
-52.8	-63	-81.4	-22.8	-9	15.8	7.2	45	113.0	37.2	99	210.2
-52.2	-62	-79.6	-22.2	-8	17.6	7.8	46	114.8	37.8	100	212.0
-52.2	-61	-77.8	-21.7	-7	19.4	8.3	47	116.6	38.3	101	213.8
-51.6	-60	-76.0	-21.1	-6	21.2	8.9	48	118.4	38.9	102	215.6
-51.1	-59	-74.2	-20.6	-5	23.0	9.4	49	120.2	39.4	103	217.4
-50.6	-58	-72.4	-20.0	-4	24.8	10.0	50	122.0	40.1	104	219.2
-49.4	-57	-70.6	-19.4	-3	26.6	10.6	51	123.8	40.6	105	221.0
-48.9	-56	-68.8	-18.9	-2	28.4	11.1	52	125.6	41.1	106	222.8
-48.3	-55	-67.0	-18.3	-1	30.2	11.7	53	127.4	41.6	107	224.6
-47.8	-54	-65.2	-17.8	0	32.0	12.2	54	129.2	42.2	108	226.4
-47.2	-53	-63.4	-17.2	1	33.8	12.8	55	131.0	42.7	109	228.2
-46.7	-52	-61.6	-16.7	2	35.6	13.3	56	132.8	43.3	110	230.0
-46.1	-51	-59.8	-16.1	3	37.4	13.9	57	134.6	43.8	111	231.8
-45.6	-50	-58.0	-15.6	4	39.2	14.4	58	136.4	44.4	112	233.6
-45.0	-49	-56.2	-15.0	5	41.0	15.0	59	138.2	45.0	113	235.4
-44.4	-48	-54.4	-14.4	6	42.8	15.6	60	140.0	45.5	114	237.2
-43.9	-47	-52.6	-13.9	7	44.6	16.1	61	141.8	46.1	115	239.0
-43.3	-46	-50.8	-13.3	8	46.4	16.7	62	143.6	46.6	116	240.8
-42.8	-45	-49.0	-12.8	9	48.2	17.2	63	145.4	47.2	117	242.6
-42.2	-44	-47.2	-12.2	10	50.0	17.8	64	147.2	47.7	118	244.4
-41.7	-43	-45.4	-11.7	11	51.8	18.3	65	149.0	48.3	119	246.2
-41.1	-42	-43.6	-11.1	12	53.6	18.9	66	150.8	48.8	120	248.0
-40.6	-41	-41.8	-10.6	13	55.4	19.4	67	152.6	49.4	121	249.8
-40.0	-40	-40.0	-10.0	14	57.2	20.0	68	154.4	50.0	122	251.6
-39.4	-39	-38.2	-9.4	15	59.0	20.6	69	156.2	50.5	123	253.4
-38.9	-38	-36.4	-8.9	16	60.8	21.1	70	158.0	51.1	124	255.2
-38.3	-37	-34.6	-8.3	17	62.6	21.7	71	159.8	51.6	125	257.0
-37.8	-36	-32.8	-7.8	18	64.4	22.2	72	161.6	52.2	126	258.8
-37.2	-35	-31.0	-7.2	19	66.2	22.8	73	163.4	52.7	127	260.6
-36.7	-34	-29.2	-6.7	20	68.0	23.3	74	165.2	53.3	128	262.4
-36.1	-33	-27.4	-6.1	21	69.8	23.9	75	167.0	53.8	129	264.2
-35.6	-32	-25.6	-5.6	22	71.6	24.4	76	168.8	54.4	130	266.0
-35.0	-31	-23.8	-5.0	23	73.4	25.0	77	170.6	54.9	131	267.8
-34.4	-30	-22.0	-4.4	24	75.2	25.6	78	172.4	55.5	132	269.6
-33.9	-29	-20.2	-3.9	25	77.0	26.1	79	174.2	56.1	133	271.4
-33.3	-28	-18.4	-3.3	26	78.8	26.7	80	176.0	56.6	134	273.2
-32.8	-27	-16.6	-2.8	27	80.6	27.2	81	177.8	57.2	135	275.0
-32.2	-26	-14.8	-2.2	28	82.4	27.8	82	179.6	57.7	136	276.8
-31.7	-25	-13.4	-1.7	29	84.2	28.3	83	181.4	58.3	137	278.6
-31.1	-24	-11.2	-1.1	30	86.0	28.9	84	183.2	58.8	138	280.4
-30.6	-23	-9.4	-0.6	31	87.8	29.4	85	185.0	59.4	139	282.2
									60.0	140	284.0

Example: 50°F = 10.0°C
40°C = 104.0°F

Table B-8. Temperature Conversion Chart

Altitude	Temp (°C)
Sea Level	+ 15° C
3,000 feet	+ 9° C
5,000 feet	+ 5° C
7,000 feet	+ 1° C
9,000 feet	- 3° C
11,000 feet	- 7° C
13,000 feet	- 11° C
15,000 feet	- 15° C
17,000 feet	- 19° C
19,000 feet	- 23° C
21,000 feet	- 27° C
23,000 feet	- 31° C
25,000 feet	- 35° C
27,000 feet	- 38° C
29,000 feet	- 42° C
31,000 feet	- 46° C
33,000 feet	- 50° C
35,000 feet	- 54° C
37,000 feet and higher	- 57° C

Table B-9. Standard Temperature versus Altitude Chart

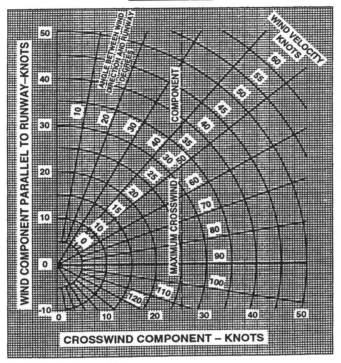

Table B-10. Wind Component Chart

List of Mental Math Formulas

Reciprocal Headings

When Init Hdg <180°, Init Hdg +200° − 20° =
 Recip Hdg°

When Init Hdg >180°, Init Hdg −200° + 20° =
 Recip Hdg°

Hydroplaning

$V_{HP} = 9\sqrt{\text{Tire Pressure}}$

Temperature Conversions

$°F = ([9/5] \times °C) + 32$

$°C = (°F − 32) \times (5/9)$

Temperature Lapse Rate

ISA Temp Lapse Rate = 2°C (or 3.5°F) per 1,000'

Moist Temp Lapse Rate = 2.5°C (or 4.5°F) per 1,000'

Pressure Altitude

PA = Indicated Altitude ± Altimeter
   Setting Correction

Altimeter Setting Correction =
   (QNH − 29.92) × 1,000'

Drift Angle

Drift Angle =
   (Crosswind Component) × 60 ÷ TAS

Avgas

Total pounds Avgas =
   (# Gallons) × (6.0 lbs per gallon)

Total gallons Avgas =
   (Pounds Avgas) ÷ (6.0 lbs per gallon),

or

Total gallons Avgas = [(Pounds Avgas) × (1⅔)] ÷ 10

Jet A

Total pounds Jet A =
   (# Gallons) × (6.7 lbs per gallon)

Total gallons Jet A =
   (Pounds Jet A) ÷ (6.7 lbs per gallon),

or

Total gallons Jet A = [(Pounds Jet A) × (1½)] ÷ 10

Fuel Dumping

Fuel Dumped ÷ Dump Rate = Time

Dump Rate × Time = Fuel Dumped

60-to-1 Rule

Number of Radials per mile = 60 ÷ DME

Width of 1 degree (NM) = DME ÷ 60

Turn Radius

Turn Radius (NM) = True Airspeed [TAS] ÷ 200

$$\text{Turn Radius (NM)} = \frac{\text{TAS} \times 1\%}{2}$$

Turn Radius (NM) = (Mach Number × 10) − 2

SRT Bank Angle

Bank Angle (SRT) = (TAS ÷ 10) × 1.5

True Airspeed

KTAS = KIAS + (KIAS × [Altitude {in 1,000s} × 2%])

Time-Speed-Distance

Ground Speed × Time = Distance

Enroute Descents

Top of Descent (TOD) =
   [Altitude (in 1,000s) to lose × 3] + (Slowdown
   distance) + (Crossing restriction DME)

TOD =
   [Altitude (in 1,000s) to lose ÷ 3] × 10 + (Slowdown Distance) + (Crossing restriction DME)

Visual Descent Points

DME = (HAT ÷ 300) + (DME @ Runway Threshold)

Time = (Timing for Approach) − (HAT ÷ 10)

Glossary of Acronyms and Terms

AGL above ground level

AIM Aeronautical Information Manual

ANDS An acronym to remember turning characteristics with reference to the magnetic compass—Accelerate North, Decelerate South.

ATC Air Traffic Control

ATIS Automatic Terminal Information Service

C Centigrade/Celsius

C_L coefficient of lift

DME distance measuring equipment

ETE estimated time enroute

F Fahrenheit

FAF final approach fix

FAR Federal Aviation Regulations

FL flight level

FPM feet per minute

G force of gravity

GAL gallons

GS ground speed

GS glideslope

HAT height above touchdown

HG mercury

HW headwind

Hydroplaning A condition resulting from the action of a tire squeezing water from between the tire tread and the surface upon which the tire is rolling. This squeezing action generates water pressure that can lift portions of the tire off the surface and reduce the amount of friction the tire can develop. The loss can be partial or complete. There are three types of hydroplaning: dynamic, reverted rubber, and viscous.

IAS indicated airspeed

ICAO International Civil Aviation Organization

IFR instrument flight rules

ILS instrument landing system

IN inches

ISA (International standard atmosphere). A standard atmosphere. The value of temperature, pressure, and density at sea level in the standard atmosphere: temperature = 59°F (15°C), pressure = 29.92 in. Hg, density = 0.0023769 slugs/ft3.

K knots (nautical miles per hour)

KIAS knots indicated airspeed

KM kilometer

KPH kilometers per hour

KTAS knots true airspeed

L left

L lift

LBS pounds

LOC localizer

M mach number

MAP missed approach point

MDA minimum descent altitude

METAR aviation routine weather report

MIN minute

MPH miles per hour

MPM miles per minute

MPS meters per second

MSL mean sea level

NM nautical mile

NOTAM Notices to Airmen

PAPI precision approach path indicator

PDP planned descent point

PIREP pilot reports

PPM pounds per minute

PSI pounds per square inch

QNE The barometric pressure used for the standard altimeter setting (29.92 inches Hg).

QNH local altimeter setting (above MSL)

R right

Radial A line of radio bearing radiating outward from a very-high-frequency omnirange (VOR) navigation facility. There are 360 radials radiating out from each VOR, and each radial is named for the number of degrees clockwise from magnetic north that the radial leaves the facility.

Reciprocal opposite

ρ (Rho) air density

RVR runway visual range

S wing surface area

SM statute mile

SRT standard rate turn

TAF terminal aerodrome forecast

TAS true airspeed

TERPs United States Standard for Terminal Instrument Procedures

TOD top of descent

TW tailwind

UNOS Undershoot North, Overshoot South

V velocity

VASI visual approach slope indicator

VDP visual descent point

VFR visual flight rules

VHF   very high frequency

V_{HP} hydroplaning speed

VMC   visual meteorological conditions

VOR   VHF omnidirectional range

Index

About the Author

Mental Math for Pilots is the second book written by Ron McElroy. After writing his first book, *Airline Pilot Technical Interviews*, it became apparent that many pilots were struggling to solve math problems in the cockpit without the use of calculators or even a pencil and paper to help. Ron drew upon his many years as an instructor in the classroom and cockpit to capture these mental math struggles and provide solutions that are simple to use and practical for airborne contingencies. Many of his problem-solving techniques are borrowed from his own experiences ranging from computing ETAs for position reports to asking "What's it doing now?" of the current generation of glass cockpit airplanes.

Since 1976, Ron has flown in nearly every area of aviation possible. He's been an Air Force test pilot at Edwards Air Force Base; a flight and ground instructor for the Air Force, several FBOs, and an aviation college; a charter pilot; a skydive pilot;

photo and chase plane pilot; simulator instructor and line pilot for two airlines. He is currently flying for a major airline. Ron has flown 93 types of aircraft in his career, from the Piper Cub to the Boeing 777, and the military T-38 to the C-17.

Ron has always maintained a multifaceted interest in all levels of teaching pilots about the technical aspects of their profession. Armed with his broad experience, he has provided a tremendous service to those pilots needing just a little help to start new careers as airline pilots or sharpening their skills as professionals.

In his role of coaching aspiring airline pilots, Ron has discovered that technical preparation and mental math skills are key to successful airline interviews. Ron's teaching and techniques gives all pilots a sharper edge.

Professional Aviation Series from ASA

Books/CD/Online Training

Checklist for Success: A Pilot's Guide to the Successful Airline Interview
By Cheryl Cage. Over 20,000 copies sold. Updated yearly.

Checklist for Success CD:
Virtual Interview Preparation
By Cheryl Cage. Applicants answer questions in a correct/incorrect manner and Cheryl critiques. Also, paperwork, self-evaluation. (Companion to *Checklist* book.)

Airline Pilot Technical Interviews
By Ronald McElroy. Approach plates, weather, AIM, FARs, mental math, cockpit situations to analyze.

Mental Math For Pilots
By Ronald McElroy. Mental math tips and tricks for interview and cockpit use.

Reporting Clear? A Pilot's Guide to Background Checks
By Cheryl Cage. Pre-employment background checks are an important part of the selection process. Do-it-yourself background check and reasons

why you should conduct your OWN background check prior to filling out employment applications.

The Resilient Pilot: A Pilot's Guide to Surviving, & Thriving, During Furlough

By Cheryl Cage. Motivational guidance to help find enjoyable work outside the cockpit.

PILOT E–Training Test: Mental Math

By McElroy/Cage. Gauge your mental math abilities then improve them with this online mental math study tool. To order this online study guide, visit www.cageconsulting.com

Welcome Aboard! Your Career as a Flight Attendant

By Becky S. Bock. A complete guide to understanding the job of F/A and preparing for interviews.

Your Job Search Partner

By Cheryl Cage. A ten day, opportunity producing job search planning guide.

Calm In The Face Of Conflict

By Cheryl Cage. Eleven surprisingly simple strategies for handling everyday conflicts, decisions, and problems.

Technical Flash Cards: FAR & AIM

By Ronald McElroy.

·   For more information or to order any of these books, call 1-800-ASA2FLY, or visit www.asa2fly.com